Seeking Patterns, Building Rules

Algebraic Thinking

STUDENT BOOK

TERC

Mary Jane Schmitt, Myriam Steinback,
Tricia Donovan, and Martha Merson

Bothell, WA • Chicago, IL • Columbus, OH • New York, NY

TERC
2067 Massachusetts Avenue
Cambridge, Massachusetts 02140

EMPower Research and Development Team
Principal Investigator: Myriam Steinback
Co-Principal Investigator: Mary Jane Schmitt
Research Associate: Martha Merson
Curriculum Developer: Tricia Donovan

Contributing Authors
Donna Curry
Marlene Kliman

Technical Team
Graphic Designer and Project Assistant: Juania Ashley
Production and Design Coordinator: Valerie Martin
Copyeditor: Jill Pellarin

Evaluation Team
Brett Consulting Group:
Belle Brett
Marilyn Matzko

EMPower™ was developed at TERC in Cambridge, Massachusetts. This material is based upon work supported by the National Science Foundation under award number ESI-9911410. Any opinions, findings, and conclusions or recommendations expressed in this publication are those of the authors and do not necessarily reflect the views of the National Science Foundation.

TERC is a not-for-profit education research and development organization dedicated to improving mathematics, science, and technology teaching and learning.

http://empower.terc.edu

Printed in the United States of America
3 4 5 6 7 8 9 QVS 17 16 15 14

ISBN 978-0-07662-088-3
MHID 0-07-662088-3

Contents

Introduction

Welcome to EMPower

Students using the EMPower books often find that EMPower's approach to mathematics is different from the approach found in other math books. For some students, it is new to talk about mathematics and to work on math in pairs or groups. The math in the EMPower books will help you connect the math you use in everyday life to the math you learn in your coursel.

We asked some students what they thought about EMPower's approach. We thought we would share some of their thoughts with you to help you know what to expect.

> *"It's more hands-on."*
>
> *"More interesting."*
>
> *"I use it in my life."*
>
> *"We learn to work as a team."*
>
> *"Our answers come from each other… [then] we work it out ourselves."*
>
> *"Real-life examples like shopping and money are good."*
>
> *"The lessons are interesting."*
>
> *"I can help my children with their homework."*
>
> *"It makes my brain work."*
>
> *"Math is fun."*

EMPower's goal is to make you think and to give you puzzles you will want to solve. Work hard. Work smart. Think deeply. Ask why.

Using This Book

This book is organized by lessons. Each lesson has the same format.

- The first page explains the lesson and states the purpose of the activity. Look for a question to keep in mind as you work.
- The activity page comes next. You will work on the activities in class, sometimes with a partner or in a group.
- Look for shaded boxes with additional information and ideas to help you get started if you become stuck.

- Practice pages follow the activities. These practices will make sense to you after you have done the activity. The four types of practice pages are

  Practice: provides another chance to see the math from the activity and to use new skills.

  Symbol Sense Practice: offers an opportunity to become comfortable with algebraic notation.

  Extension: presents a challenge with a more difficult problem or a new but related math idea.

  Test Practice: asks a number of multiple-choice questions and one open-ended question.

In the *Appendices* at the end of the book, there is space for you to keep track of what you have learned and to record your thoughts about how you can use the information.

- Use notes, definitions, and drawings to help you remember new words in *Vocabulary*, pages 155–157.
- Answer the *Reflections* questions after each lesson, pages 158–163.

Opening the Unit: Seeking Patterns, Building Rules

What makes something a pattern?

You are about to set out on a journey to the world of **algebra**. It is a world full of **patterns** and relationships, a world that opens doors to understanding mathematics in a way that prepares you to be a better thinker and decision-maker. Algebra also prepares you to move on to further education and training. The only items you need to pack for this journey are your knowledge of numbers and your powers of concentration and imagination.

In this opening session, you will have a chance to show some things you already know about algebra. You will be asked to describe some patterns, including a pattern of your own.

Activity 1: Algebra Mind Map

Make a Mind Map using words, numbers, pictures, or ideas that come to mind when you think of *algebra*.

Activity 2: One of My Own Patterns

1. Describe in complete sentences a pattern in your life that involves numbers.

2. Tell how the pattern plays out over time, in terms of a week, a month, or even longer. Explain how the numbers grow or decrease over time and how they help you predict the future!

My pattern:____

3. Use a different representation—a diagram, table, graph, or equation—to show the pattern you just described in sentences. Try to make a representation that someone else could look at to predict what the outcome of your pattern would be in a year or longer.

Activity 3: Initial Assessment—Patterns and Problems

Your teacher will show you some problems and ask you to check off how you feel about your ability to solve each problem.

___ Can do ____ Don't know how ____ Not sure

Seeking Patterns, Building Rules Unit Goals

What are your goals regarding the study of algebra? Review the following goals. Then think about your own goals and record them in the space provided.

- Identify patterns and predict outcomes in a variety of situations.
- Describe patterns and relationships using diagrams, words, tables, graphs, and/or equations.
- Understand how different representations are related.
- Recognize the characteristics of linear patterns.
- Use basic algebraic notation.

My Own Goals

Tips for Success

Start noticing patterns and relationships everywhere you go.

Keep your eye on the prize: What is the situation about?

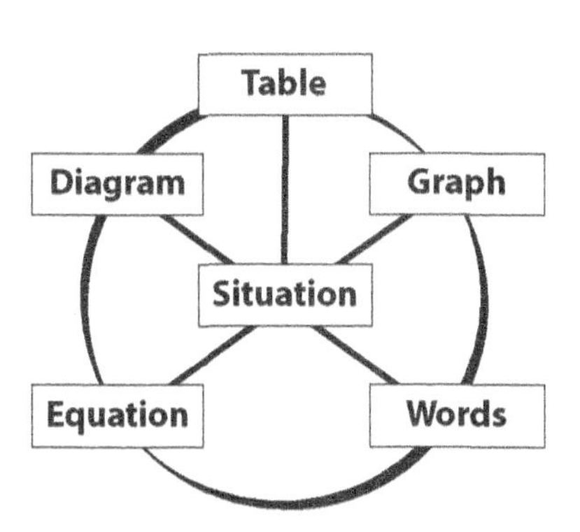

Do not get lost in the numbers, letters, and lines. Keep going back to what the situation is about—for example, what is happening to your heart rate as you age or to total earnings over time. The tables, rules, diagrams, graphs, and equations are tools that you can use to get a better idea of what is happening in the situation. The diagram at left shows these five different ways of representing a situation.

Keep connecting the diagrams, equations, tables, charts and graphs to one another.

If you have extra time after completing a table, ask yourself: "What would a graph of this look like? What's the rule in words or with an equation? And what do these ways of representing the situation tell us?

Am I done?

Do not walk away yet. Check your answers to make sure they make sense. Check your math with a calculator. Ask others whether your work makes sense to them.

LESSON 1

Guess My Rule

How can I explain the rule behind this table of numbers?

Have you ever spotted a pattern in the numbers in health charts, tax **tables**, or sports scores? Sometimes there is a rule or formula built into the chart. Many jobs now require employees to use computer spreadsheets to keep track of products and expenses. Spreadsheets are another example of tables of numbers based on formulas or rules.

In this lesson, you will look for the rules behind the numbers in tables. You will play a game with **In-Out tables** that show various patterns of numbers. You will try to figure out what the pattern is; then you will describe each pattern in words and equations.

Activity 1: Guess My Rule

Can you guess my rule?

If you put 20 in...

...out comes 40.

x	y
20	40
45	90
92	184

How to Play Guess My Rule

- Look for a pattern in the numbers in the table.
- Suggest a new pair of numbers for the table that follows the pattern and fill in any missing values.
- Write down the rule (what did you do to x to get y?):

  In words.

  In symbols as an **equation**.

Check that the rule works for every case.

Activity 2: Making Tables, Rules, and Equations

In this round of *Guess My Rule*, you and your partner(s):

- Make up a rule in words.
- Create a table to go with it. Leave one or two entries blank.
- Write an equation for your rule.

Remember: Your rule can use any operation or combination of operations.

Our rule in words:

Our table:

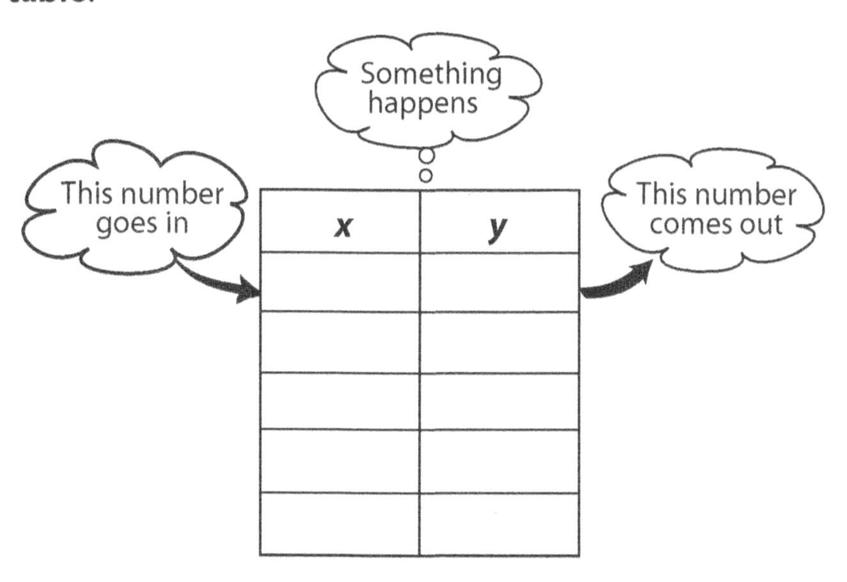

x	y

Our equation:

Practice: Guess More Rules

An In-Out table always has a rule to get from the In value (x) to the Out (y). For each of the following tables:

- Look for a pattern in the table.
- Try to figure out what the rule is to get from x to y.
- Check to see whether the rule works for every case, and write in another pair of numbers that follows the rule.
- Write the rule in words.
- Write an equation for the rule.

Example

x	y
3	15
4	20
5	25
12	60

Fill in the table.

The rule in words:

Multiply x by five to get y

The equation:

$y = 5x$

Check your rule by reading it out loud. If you follow the rule, do the pairs of numbers work?

Table 1

x	y
7	28
8	32
9	36

Fill in the table.

The rule in words:

The equation:

Table 2

x	y
3	24
5	40
6	48

Fill in the table.
The rule in words:

The equation:

Table 3

x	y
4	9
5	10
6	11

Fill in the table.
The rule in words:

The equation:

Table 4

x	y
6	5
8	7
13	12

Fill in the table.
The rule in words:

The equation:

Table 5

x	y
2	1.0
5	2.5
8	4.0

Fill in the table.
The rule in words:

The equation:

Table 6

x	y
1	4
2	7
3	10
4	13

Fill in the table.
The rule in words:

The equation:

Table 7

x	y
3	22
4	29
11	78

Fill in the table.
The rule in words:

The equation:

The rules for the next three tables have a twist.

Table 8

x	y
1	1
3	9
7	49

Fill in the table.
The rule in words:

The equation:

Table 9

x	*y*
2	6
5	27
6	38
10	102

Fill in the table.
The rule in words:

The equation:

Table 10

x	*y*
1	19
2	18
6	14
7	13

Fill in the table.
The rule in words:

The equation:

Practice: Fill In the Values

Figure out the rule for the following tables and fill in the missing values.

Example

x	*y*
60	30
56	28
42	21
100	50
200	100

1.

x	y
1	7
2	9
3	11
4	
5	

2.

x	y
12	13
11	14
16	9
4	
	20

3.

x	y
1	2
2	5
3	8
10	
	98

4.

x	y
100	5
101	6
102	7
200	
	200

5.

x	y
2	
3	8
	15
10	99
12	

6.

x	y
5	20
10	95
12	139
15	
20	

Practice: The Rule Story

Fill in the blanks using the words and numbers from the box below. Some words may be used more than once.

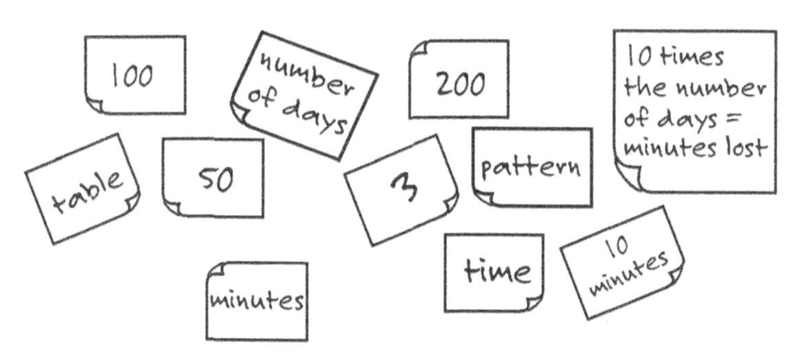

Terri managed an office where one of the bosses was always 10 minutes late for work. This annoyed Terri, so she decided to make a __________ to keep track of the accumulated lost work time. "I'm going to show James how his __________ of lateness mounts up over __________," she thought. The two things Terri recorded were the number of days James was late and the number of lost __________ that built up, or accumulated. On day one, Terri recorded that 10 minutes of work time had been lost. By the end of the workweek, her __________ showed that James had been late 10 minutes every workday and had lost a total of __________ minutes of work time.

Terri decided to confront James. "Look at this __________," she said. "You can see that in a month (20 workdays), you will have cost this company __________ lost minutes of work time. That's more than __________ hours. And at your salary, that's a lot of money!"

James didn't believe Terri. "Show me how you figured that out," he said.

Terri agreed to explain how she could predict the time James would lose because of his __________ of lateness. "See here," she said. "You take the number of __________ late each day, always the same, and multiply it by the __________ you will be late to find out the total number of minutes you will have lost over the course of a month." Just to make her point, she wrote on a piece of paper the rule for James's __________. That rule looked like this: __________

James was paid $30 per hour. Even he could figure out that at that rate of pay, his lateness cost the company __________ dollars a month. He was ashamed of himself, but proud of his manager. "Terri," he said, "I always knew you were good at math. I just didn't know *how* good."

Symbol Sense Practice: How Many Ways?

Complete each row by writing the multiplication statements in two other ways.

You can show multiplication in several ways.

2×3

$2(3)$

$2 \bullet 3$

two times three

Did you know that two letters next to each other also indicate multiplication?

$rt = d$

$r \times t = d$

$r(t) = d$

r times *t* equals *d*

Five times four		
3×42		
39 (84)		
ab		
lwh		
$5 \bullet 10$		
$24x + 3$		
$b = 100 - 6a$		
$I = prt$		

Symbol Sense Practice: Four Ways to Write Division

Division can be written in several ways.

If this is a picture of one candy bar... 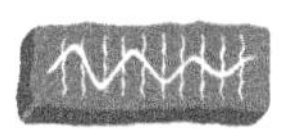

This would mean 10 candy bars divided between 2 people.

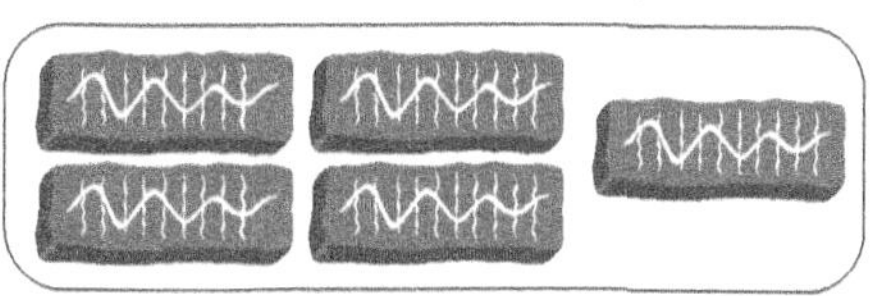

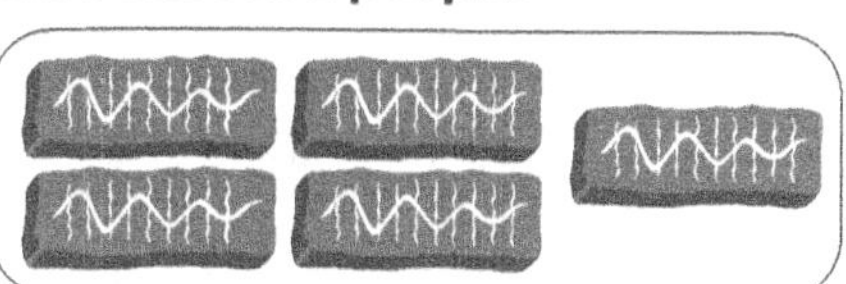

$2\overline{)10}$	$10 \div 2$	$\frac{10}{2}$	10 divided by 2

Be careful: The order of the numbers makes a difference in division.

This would mean 2 candy bars divided among 10 people.

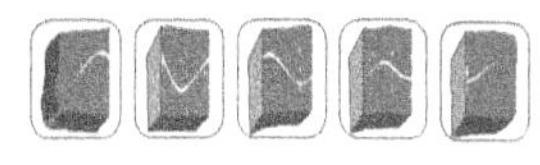

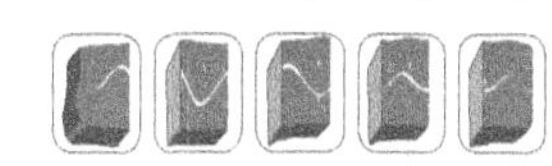

$10\overline{)2}$	$2 \div 10$	$\frac{2}{10}$	2 divided by 10

Complete each row by writing the division three other ways.

$8\overline{)56}$	$56 \div 8$	$\frac{56}{8}$	56 divided by 8
	$30 \div 5$		
$6\overline{)180}$			
	$40 \div 1$		
			2 divided by 7
		$\frac{x}{y}$	
			m divided by 10
$\$.50\overline{)\$10.00}$			

Extension: Building Tables from Rules

Make an In-Out table for each symbolic rule. Check to be sure the numbers fit the rule.

$y = \frac{1}{4}x$

x	y

$y = 2x + 0.75$

x	y

$y = \frac{1}{2}x - 1$

x	y

$y = 1.8x$

x	y

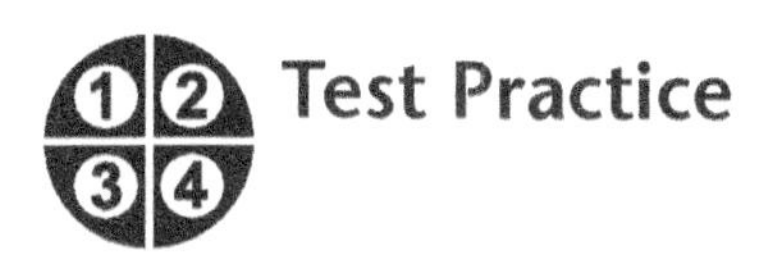

Test Practice

1. Which of the following is another way to express $x + x + x$?

   (1) $3x$

   (2) $\frac{x}{3}$

   (3) $3 + x$

   (4) x

   (5) $3 - x$

2. Which of the following rules describes the pattern in the following table?

x	y
36	4
63	7
108	12

   (1) Multiply x by seven to get y.

   (2) Divide 9 by x to get y.

   (3) Multiply x by nine to get y.

   (4) Divide x by nine to get y.

   (5) Add 56 to x to get y.

3. What is the missing number in the following In-Out table?

x	y
8	
9	19
15	31
2.5	6

   (1) 16

   (2) 17

   (3) 18

   (4) 19

   (5) 20

4. Which equation below represents the pattern in the following table?

x	y
12	120
23	230
0.25	2.5

   (1) $100 + x = y$

   (2) $y = 10x$

   (3) $\frac{x}{10} = y$

   (4) $y = x(12)$

   (5) $x + 10 = y$

5. Which of the rules below describes the pattern in the following table?

x	y
14	27
20	39
101	201

(1) Multiply x by two to get y.

(2) Divide x by two to get y.

(3) Multiply x by two, then add one to get y.

(4) Divide x by two, then subtract one to get y.

(5) Multiply x by two, then subtract one to get y.

6. What is the missing number in the following table?

x	y
7	3
28	24
30	

LESSON 2

Banquet Tables

How many tables will I need for a crowd of any specific size?

Men and women in the catering business need to be masters at planning and predicting. A good caterer can accurately forecast how much food to cook, how much time to plan for preparation, and how many banquet tables to set up, depending on the number of people expected.

In this lesson, you will build some banquet table arrangements to help you look for patterns. You will also use diagrams and a well-organized table of numbers to help you see the relationships between the variables in a problem—in the first case, the number of square tables and the number of people seated at them. You will describe the pattern and write a rule for it. You will discover that no matter how big the crowd, you can figure out how many banquet tables will be needed, once you have developed a rule.

Activity: Banquet Tables

Arrangement 1 | Arrangement 2 | Arrangement 3

Consider this banquet table pattern. Think of Arrangement 1 as a small, square table that can seat four people. The second arrangement shows two of the square tables pushed together to make a longer banquet table, and the third arrangement shows three square tables pushed together to make an even longer table.

Task 1

Given the number of square tables, find the number of people who could be seated around them.

1. How many people would fit around each arrangement of banquet tables?

   Arrangement 2 _____ Arrangement 3 _____

2. Draw the next table arrangement in this pattern, and write down how many people could be seated at it.

3. How many people could sit around an arrangement of 10 tables pushed together?

   a. 40

   b. 22

   c. 10

   d. None of these

   Explain your answer.

4. Explain how you would figure out the number of people who could fit around an arrangement of 100 tables pushed together.

5. Make a table to keep track of the number of banquet table arrangements and the number of people who could be seated for each arrangement. Record the information you have gathered so far.

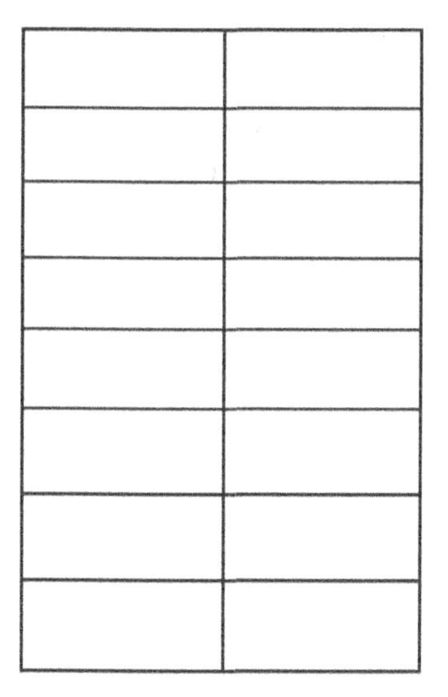

6. What is the rule for finding out how many people you would be able to seat at *any* number of square table arrangements?

   In words:

   As an equation:

7. Does your rule work when you apply it to the numbers in your table? Give at least one example that proves this is so.

Task 2

Given the number of people, find the number of square tables need to seat them.

1. How many square tables would be needed to seat 20 people?

   a. 42

   b. 80

   c. 9

   d. None of these

   Explain your answer.

2. Explain how you would figure out the number of square tables needed to seat 100 people.

3. What is a rule for finding out how many square tables you would need for any number of people?

4. Does your rule work when you apply it to the numbers in your table? Give at least one example that proves this.

5. How does the rule for finding the number of people compare with the rule for finding the number of tables?

Problem Response

1. What did you learn by working on this problem?

2. What are you still wondering about?

Practice: Toothpick Row Houses

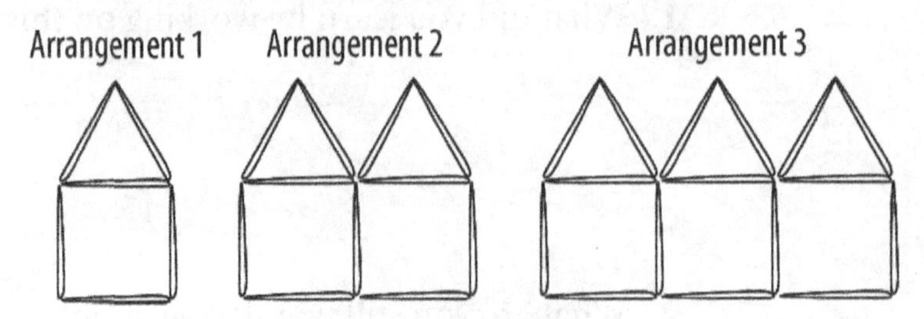

Consider this pattern. Think of the first picture as a house built of six toothpicks. The second picture is of two houses connected in a row, and the third arrangement is of three houses connected to make an even longer row.

1. How many toothpicks does it take to build the second arrangement?

2. How many toothpicks are needed for the third arrangement?

3. What would the fourth arrangement look like?

4. Create a table showing arrangements 1, 2, 3, and 4 (n) and the number of toothpicks (t) needed for each arrangement.

5. How many toothpicks would it take to build an arrangement of 10 row houses? Explain your reasoning, using words or diagrams or both.

6. Explain how you would figure out the number of toothpicks needed for an arrangement of 100 row houses.

7. The algebraic question: How many toothpicks will be needed for an arrangement of *any* number of houses?

   a. Write a rule that would allow anyone to predict the number of toothpicks needed for any specific number of houses.

   b. Give a few examples to show that your rule works.

8. If you had a box of 126 toothpicks, how many houses could you build? Write a rule that would allow anyone to predict the number of houses you could build with 126 toothpicks.

Symbol Sense Practice: Rules of Order

My Dear Aunt Sally

When you read text, you read from left to right. In math notation, you do not always proceed from left to right. Sometimes you start at the middle or the end of an expression. Always start with the multiplication and division, then do the addition and subtraction. One way to remember this is with the phrase "**M**y **D**ear **A**unt **S**ally," where the initial letters of each word stand for multiplication, division, addition, and subtraction.

Example:

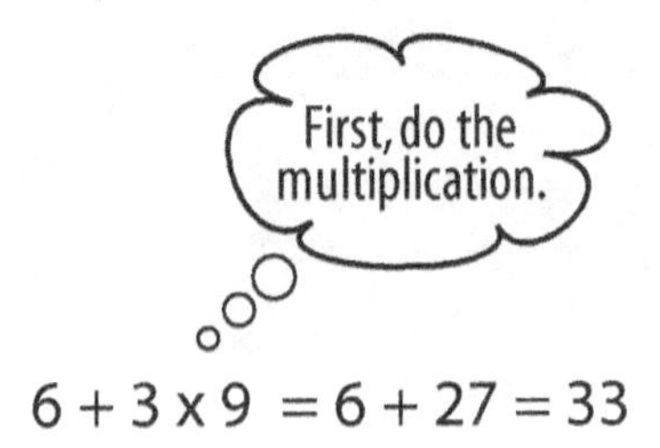

$$6 + 3 \times 9 = 6 + 27 = 33$$

See how quickly you can solve each problem below.

1. $5 - 4 + \frac{6}{2} =$
2. $120 \times 4 + 3 - 2 =$
3. $100(62) - 30(62) =$
4. $\frac{100}{2} - \frac{50}{5} =$
5. $6 + 4(3) + 12 =$
6. $6 + \frac{4(3)}{12} =$

Parentheses

In mathematical notation, parentheses signal where to focus first. Evaluate each expression, paying attention to the math operations inside the parentheses first. Then perform any remaining multiplication and division before doing the addition and subtraction.

Example:

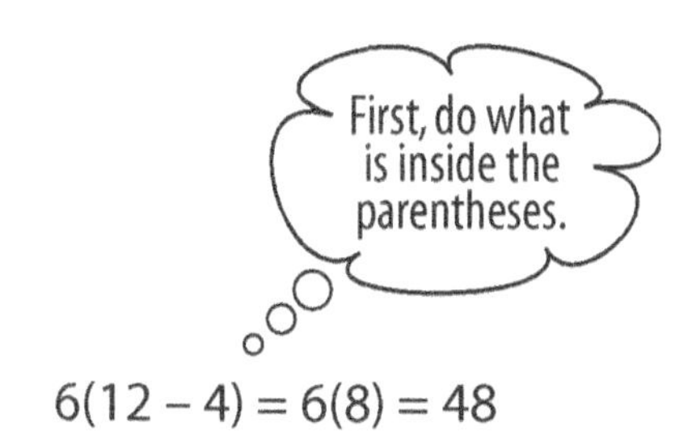

$$6(12 - 4) = 6(8) = 48$$

1. $10(1 + 20) =$
2. $(4 + 3) \times (2 + 5) =$
3. $(9 + 8 - 7)\ (6 + 5 - 3) - (2 + 1) =$
4. $10(6 - 1) + 5 =$
5. $10(6) - 1 + 5 =$
6. $10(6 - 1 + 5) =$
7. $10(6) - (1 + 5) =$

You Choose the Operations

Insert "+" or "–" signs and "( )," if necessary, to make each statement an equation.

Example: 8 7 6 5 = 14

8 + (7 − 6) + 5 = 14

1. 4 3 4 1 = 6

2. 4 3 4 1 = 3

3. 4 3 4 1 = 4

Extension: The Importance of Order

1. Write the equation and make a table for each two-step rule.

Rule 1: To find *y*, multiply *x* by 5, then add 2.

x	*y*

Equation:

Rule 2: To find *y*, add 2 to *x*, then multiply by 5.

x	*y*

Equation:

Rule 3: Multiply *x* by 5, then add 10.

x	*y*

Equation:

2. Are the three rules the same or different? Explain why.

Test Practice

1. Choose the best answer to describe the shape that comes next in this pattern.

○□△○○□□△

(1) Circle
(2) Triangle
(3) Rectangle
(4) Square
(5) None of the above

2. Which equation describes this person's jelly bean-eating pattern (rate of consumption)?

Days (x)	1	2	3	4	5
Jelly beans (y)	20	40	60	80	100

(1) $x + 20$
(2) $y + 20 = x$
(3) $20x = y$
(4) $x = 20y$
(5) $y = \frac{20}{x}$

3. Part of a tax table is found. What do you think the next entry would be?

Federal Tax Table

Price	Tax
$12.00	$.72
$13.00	$.78
$14.00	$.84
$15.00	

(1) $0.60
(2) $0.85
(3) $0.90
(4) $15.90
(5) $16.00

4. Which expression has the greatest value?

(1) $1 + 2 + 3 + 4$
(2) $1(2)(3)(4)$
(3) $(1 + 2)(3 + 4)$
(4) $1 + 2(3) + 4$
(5) $1(2) + 3(4)$

5. What is the value of $\frac{5(8-2)}{5(8-2)}$?

(1) 0
(2) 1
(3) 6
(4) 30
(5) 38

6. How many blocks will the next arrangement have?

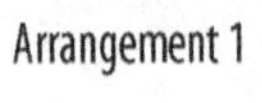

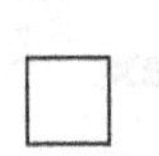

Arrangement 2

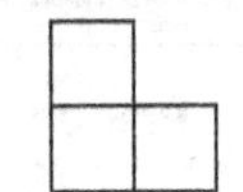

Arrangement 3

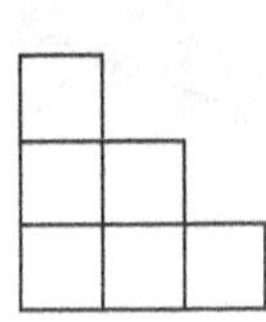

Arrangement 4

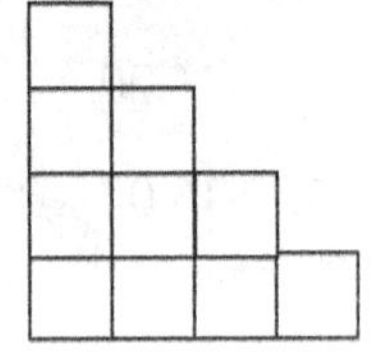

LESSON 3

Body at Work—Tables and Rules

What is your heart rate at rest?

Scientists, business people, and health professionals often use tables to organize information to look for patterns. These patterns sometimes involve *rates*, such as pay rates, interest rates, or growth rates.

This lesson gives you a chance to look for patterns of heartbeat rates and rates at which you burn calories during different activities. You will use tables to organize the information and detect the patterns. By the end of the lesson, you will be able to use the tables to describe patterns as rules both in words and in algebraic notation.

Activity 1: Heart Rates at Rest

Part 1

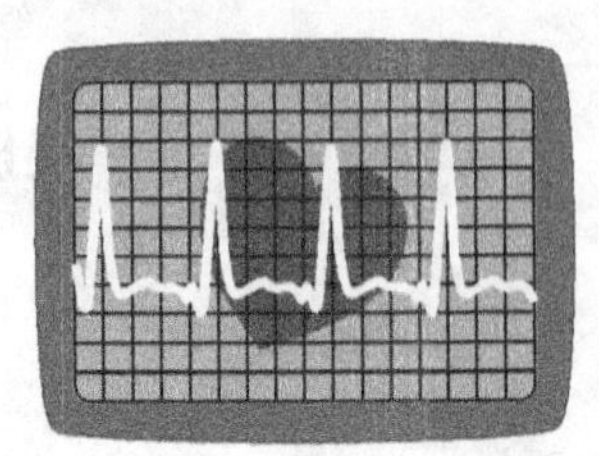

Everyone's heart throbs in a fairly rhythmic pattern. We check that pattern when we take a pulse by counting the number of heartbeats in a minute. Usually, when nurses take your pulse, they do so for less than a minute, often for 15 seconds. Take the pulses of several people for 15 seconds. Record the information in the table below. Then figure the number of heartbeats per minute for each person.

Name	Beats per 15 Seconds	Beats per Minute
A baby	30	
A feverish adult	35	

Write the rule you used for finding the number of heartbeats per minute when you knew the number of beats in 15 seconds.

Rule in words:

Rule as an algebraic equation:

Part 2

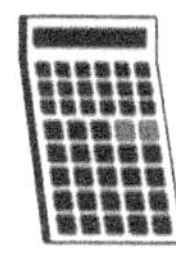

Now complete your *personal* heart-rate-at-rest table. Enter your pulse (number of heartbeats in one minute) in the first row. Then fill in the missing values based on the same rate.

Personal Heart Rate Table

Time in Minutes (*M*)	My Total Number of Heartbeats (*B*)
1	
5	
10	
30	
60 (an hour)	
1,440 (a day)	

1. Write a rule for finding the total number of times your heart *beats* if you know the number of *minutes* it has been beating.

   Rule in words:

   Rule as an algebraic equation:

Always remember to check a rule by entering the numbers in the table. A rule has to work with *all* the entries.

2. Write a rule for finding the number of *minutes* your heart beats if you know the total number of *heartbeats.*

   Rule in words:

   Rule as an algebraic equation:

3. About how long would it take for your heart to beat 1,000,000 times? How do you know?

Activity 2: How Many Calories Am I Burning?

Many people pay close attention to calories these days. There are two ways to think about calories:

We put calories into our bodies in the form of food.

We burn calories at different rates, depending on what we do and how much time we spend doing it.

In the following activity, you will consider various activities and their rates for burning calories.

- Look for a pattern in the table, and fill in the missing information. In each table, the rate for burning calories remains constant.
- Write a rule that can be used to determine the total calories burned in any number of minutes for each type of activity.

For the following tables, numbers are approximate, based upon an imaginary 5′8″, 190 lb. woman (source: http://www.caloriesperhour.com).

Minutes Jogging	Calories Burned
1	
15	150
30	
45	450
60	
100	

What patterns do you see?

The rule in words for finding calories burned while jogging:

The rule as an algebraic equation:

When writing the rule in algebraic notation, be clear about what each letter stands for. Let the reader know the letter you are using to represent calories and the letter you are using to represent minutes.

Minutes Cleaning House	Calories Burned
1	
15	75
30	
45	
60	300
100	

What patterns do you see?

The rule in words for finding calories burned while cleaning house:

The rule as an algebraic equation:

Minutes Running Up Stairs	Calories Burned
1	
15	300
30	600
45	
60	
100	

What patterns do you see?

The rule in words for finding calories burned while running up stairs:

The rule as an algebraic equation:

Minutes Sitting (Reading or Watching TV)	Calories Burned
1	
15	
30	30
45	
60	60
100	

What patterns do you see?

The rule in words for finding calories burned while sitting:

The rule as an algebraic equation:

Use the information from the four calorie-burning tables you have just completed to answer the following questions:

1. How long would you have to watch TV to burn the same number of calories as you would in a half-hour of jogging?

   a. How can you use the tables to arrive at the solution?

   b. How can you use the equations to arrive at the solution?

2. You have to burn 3,500 calories to lose a pound of fat. Invent three exercise plans for burning a pound of fat. Combine all three activities in each plan you create.

Three Ways to Burn a Pound of Fat (3,500 calories)

	Jogging		Cleaning House		Running Up Stairs		Total	
	Min.	Cal.	Min.	Cal.	Min.	Cal.	Min.	Cal.
Plan 1	180		100					3,500
Plan 2								3,500
Plan 3								3,500

Write a rule that tells, in general, how to make a plan to burn 3,500 calories. What do you add? What do you multiply?

Practice: Say It in Words and Fill in the Tables

The expressions below relate to rates of measurement that are fairly common in everyday life.

1. Write in words what each one says.
2. Fill in a table with some entries based on that rate.
3. Describe a situation where you might use that rate.

Example: 60 mph

In words: This means you travel 60 miles in one hour.

A table:

Miles	Hours
60	1
120	2
240	4
480	8

A possible situation: Driving a car on the highway

1. \$0.15/minute

In words:

A table:

A possible situation:

2. 2,500 calories/day

In words:

A table:

A possible situation:

Practice: Driving at 50 Miles per Hour

1. A woman is driving 50 miles per hour. What does that mean?

2. Does she have to drive for one entire hour to go 50 miles per hour? Explain.

3. How far does she go in

   a. 1 hour? ______

   b. 2 hours? ______

   c. Half an hour? ______

   d. 10 hours? ______

4. How long does it take her to go

   a. 50 miles? ______

   b. 100 miles? ______

   c. 25 miles? ______

   d. 500 miles? ______

5. Explain your strategies for figuring out the answers to Problems 3 and 4.

6. Put the values from Problems 3 and 4 in the table below. Be sure all your times and distances are in the same units of measurement.

Driving at 50 mph

Time (t)	Distance (d)

7. Circle all of the equations that correspond to your table:

$t = 50d$ $\qquad$ $d = 50t$ $\qquad$ $t = \frac{d}{50}$

$d = \frac{50}{t}$ $\qquad$ $d = \frac{t}{50}$ $\qquad$ $t = \frac{50}{d}$

Symbol Sense Practice: Equations ⟷ Words

Every algebraic equation can be translated into a simple sentence. Some examples are listed below:

Algebraic Equation		A Simple Sentence
$7x = y$	⟷	Multiply seven by x to find y.
$6x - 2 = y$	⟷	Six times x minus two equals y.
$y = x/4$	⟷	*y* equals *x* divided by four.

Translate each algebraic equation below into a simple sentence.

1. $x + 9 = y$ ______________________________

2. $10x + 20 = y$ ______________________________

3. $\frac{x}{8} + 15 = y$ ______________________________

4. $y = \frac{1}{2}x + 1$ ______________________________

Now write an algebraic equation for each sentence.

5. ________________________ *y* equals five multiplied by *x*.

6. ________________________ Double *x* to find *y*.

7. ________________________ Multiple *x* by ten, then add five to equal *y*.

8. ________________________ Find *y* by subtracting four from *x*.

Make up two of your own equations and matching sentences.

9. ________________________ ________________________

10. ________________________ ________________________

Symbol Sense Practice: Substituting for x

In math, a rule of order is to perform multiplication before addition, no matter where they occur. So …

$$5 + 2(100) = 5 + 200$$
$$5 + 2(100) \neq 7(100)$$

In the following equations, solve for y in three cases: when $x = 0$, when $x = 10$, and when $x = 100$. When you have addition and multiplication in the same equation, perform the multiplication first *unless* there are parentheses. If there are parentheses, do the math inside them first.

Substitution Values for x

Original Equations	$x = 0$	$x = 10$	$x = 100$
1. $y = 2x + 35$	$y =$	$y =$	$y =$
2. $y = 15 + 3x$	$y =$	$y =$	$y =$
3. $y = 5 + 7x$	$y =$	$y =$	$y =$
4. $y = 2(x + 2)$	$y =$	$y =$	$y =$
5. $y = (0.25 + 0.75)x$	$y =$	$y =$	$y =$
6. $y = 1{,}000 + 25x$	$y =$	$y =$	$y =$
7. $y = \frac{x}{2} + 90$	$y =$	$y =$	$y =$

Extension: A Friendly Reunion

1. Three friends who live quite a distance from one another planned a reunion. They picked a central meeting spot that seemed fair, 180 miles away from each person. Each of the friends has a car, but the cars are not in the same working condition. The sports car driver can drive at an average speed of 90 mph. The minivan driver figures she can go 60 mph, and the driver with the old pick-up truck will only be able to drive at 45 mph. (Highway speed limits are ignored in this problem).

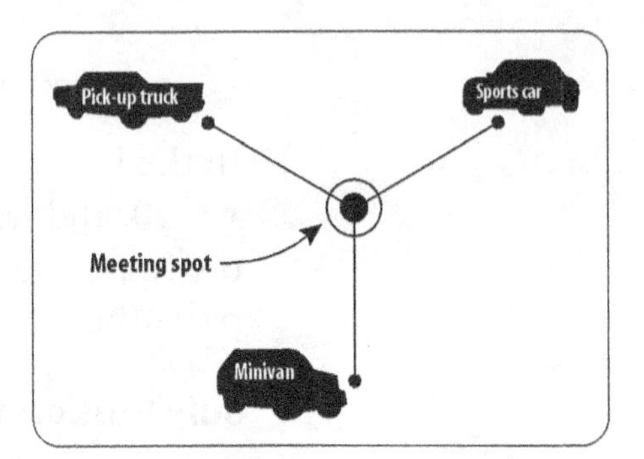

   a. Fill in three tables, one for each driver. Show at least five entries for time and distance.

Pick-up Truck

Time (in hours)	Distance (in miles)
1	
2	
3	

Sports Car

Time (in hours)	Distance (in miles)
1	
2	
3	

Minivan

Time (in hours)	Distance (in miles)
1	
2	
3	

b. If they want to meet at noon, what is the latest time each driver should plan to leave home? Use your tables and the sketch to support your answer.

2. What if some things went wrong? The three drivers all started out from their homes as planned, but halfway there, the sports car driver discovered she forgot her purse and had to go back home to get it. The minivan driver had to detour because of road construction, which meant she had to go 60 miles out of her way. The driver of the old pick-up truck had no problems. How did this affect their meeting time? Use your tables and the sketch to support your conclusion about when each person actually arrived under these new circumstances.

Test Practice

Use this table for Problems 1 and 2.

Cost of Item	Sales Tax
$20	$1.00
$30	$1.50
$40	$2.00
$50	$2.50

1. Which of the following could be the rule to find the sales tax when you know the cost of an item?

(1) Subtract $19 from the item cost.

(2) Divide the cost of the item in half.

(3) Divide the cost of the item by ten.

(4) Multiply the cost of the item by 10.

(5) Divide the cost of the item by 20.

2. Which of the following could be the rule to find the cost of the item when you know the sales tax? Let "*C*" stand for item cost and let "*t*" stand for the sales tax amount.

(1) $C = \frac{t}{20}$

(2) $C = 20(t)$

(3) $C = \frac{t}{2}$

(4) $C = 2(t)$

(5) $C = 0.5(t)$

Use the following sequence for Problems 3 and 4.

6, 11, 16, 21, ...

3. What is the eighth number in the sequence?

(1) 26

(2) 41

(3) 46

(4) 51

(5) 56

4. What digit would the 75th number in the sequence end in?

(1) 0

(2) 1

(3) 5

(4) 6

(5) 7

5. Look at the pattern in the table below. What reasonable conclusion can you make based on the information?

NutriStrategy = Calories Burned During Exercise

Activity (1 hour)	130 lb.	155 lb.	190 lb.
Running, wheeling, general	177	211	259
Sailing, boat/board, windsurfing, general	177	211	259
Sailing, in competition	295	352	431
Scrubbing floors, on hands and knees	325	387	474
Shoveling snow, by hand	354	422	518
Shuffleboard, lawn bowling	177	211	259
Sitting–playing with child(ren)–light	148	176	216
Skateboarding	295	352	431
Skating, Ice, 9 mph or less	325	387	474
Skating, Ice, general	413	493	604
Skating, Ice, rapidly, > 9 mph	531	633	776
Skating, Ice, speed, competitive	885	1056	1294
Skating, roller	413	493	604
Ski jumping (climb up carrying skis)	413	493	604
Ski machine, general	561	669	819
Skiing, cross-country, > 8 mph, racing	826	985	1208
Skiing, cross-country, moderate effort	472	563	690
Skiing, cross-country, slow or light effort	413	493	604
Skiing, cross-country, uphill, maximum effort	974	1161	1423
Skiing, cross-country, vigorous effort	531	633	776

Source: http://www.nutristrategy.com/activitylist4.htm

(1) The less you weigh, the more calories you burn doing the same amount of exercise.

(2) For any amount of exercise, the number of calories you burn increases as your weight increases.

(3) The older you are, the harder it is to lose weight.

(4) If you double your speed, you double your calories burned.

(5) The number of calories burned does not depend upon your weight.

6. Continue this pattern: What is the *fifth* number?

100, 50, 25, _____, _____, _____

LESSON **4**

Body at Work— Graphing the Information

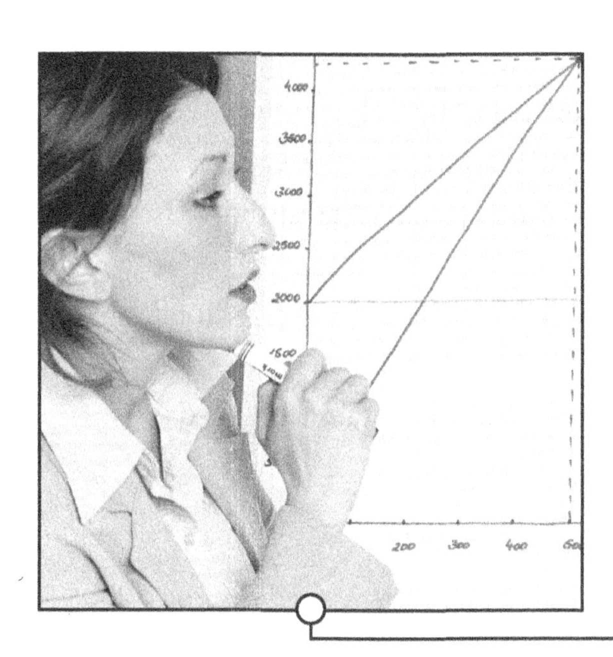

Where is the pattern in the graph?

Graphs are everywhere: in newspapers and magazines, in work situations, and on tests. Being able to read and create graphs can help you notice patterns and rules from another viewpoint.

In this lesson, you will focus on the various parts of a graph, and you will see how the patterns from *Lesson 3* can be represented with graphs, as well as with tables and rules. All of these representations of a situation are connected!

Activity: Four Graphs for Calories Burned

Tables are one way to see patterns; graphs are another. Here you will use a graph to see patterns in a new way.

> Every pair of numbers from the In-Out table can be graphed as one point on the graph.

Directions:

1. Use the grid paper on the next page.
2. On one set of axes, graph each of the four exercises from *Lesson 3* (jogging, cleaning house, running up stairs, and sitting). Use the information from the tables.
3. Mark the x-axis "Number of Minutes" and the y-axis "Total Number of Calories Burned."
4. Decide on the size of the intervals for the x-axis and y-axis.
5. Label at least three points on each graph.
6. Label each graph with its algebraic equation.

Compare the graphs of each exercise. What is the same? What is different?

Similarities:

Differences:

Four Graphs for Calories Burned

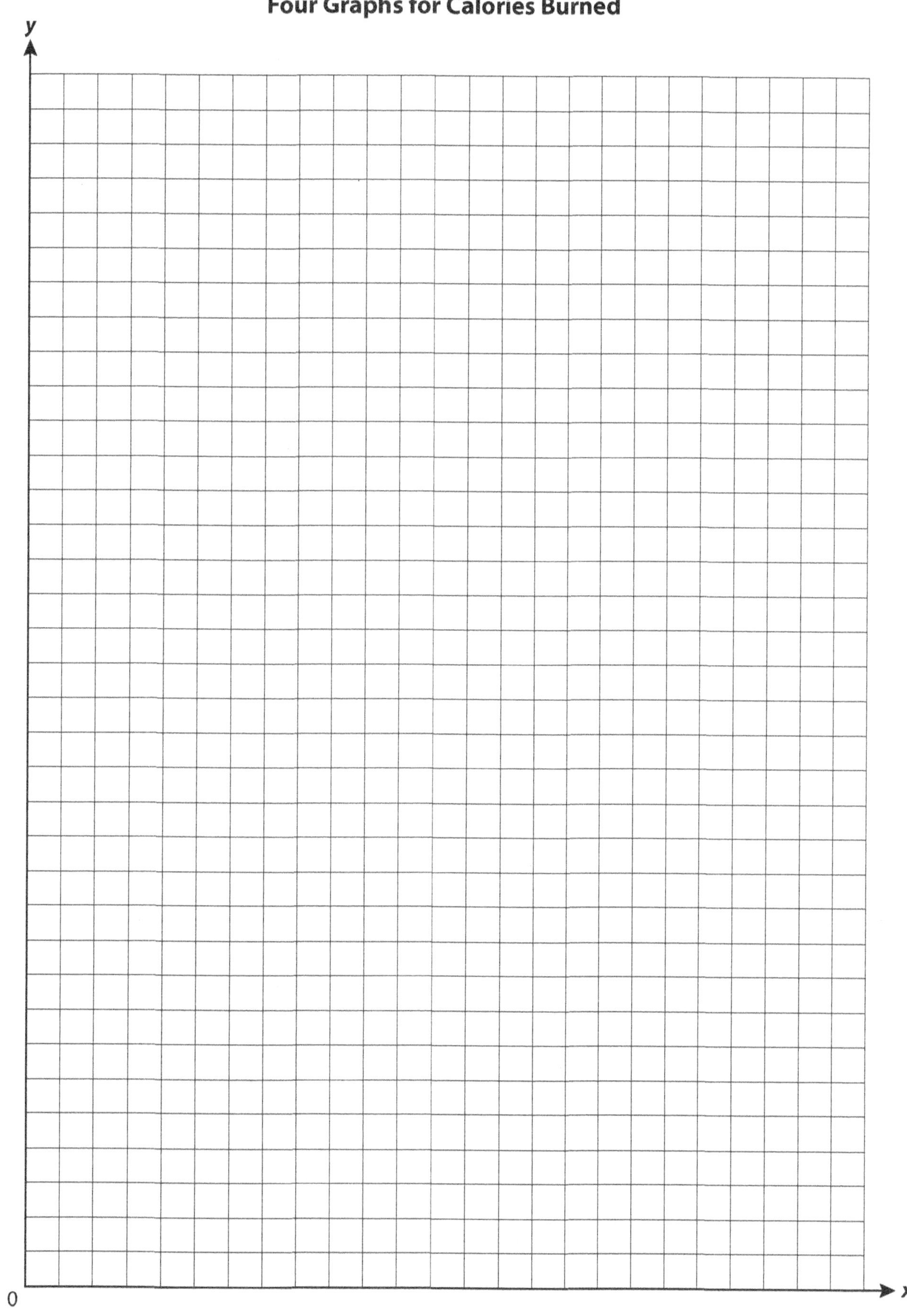

A Closer Look at Four Graphs for Calories Burned

1. Where do the four lines intersect (meet and touch)? Why do the four lines intersect at this point?

2. Which exercise graph has the steepest line? Why is the line on this graph steeper than the lines on the other graphs?

3. Which graph has the flattest line? Why is the line on this graph flatter than the lines on the other graphs?

4. Use the graphs to solve these problems. For Parts a and b, state where you found the answer on the graph.

   a. How long do you need to clean house to burn the same number of calories as you would jogging for an hour?

   b. What is the difference between the amount of calories burned in 45 minutes of sitting and in 45 minutes of running up stairs?

   c. Will the point (70, 1,400) fall on the "Running Up Stairs" graph? How do you know?

5. Shoveling snow uses up about eight calories per minute. Describe what the graph of that exercise would look like. Where will it be in relation to the other graphs? How steep will its line be?

6. Make a table for shoveling snow, showing various times and calories burned.

7. Write a rule in words and/or symbols for shoveling snow.

Practice: Setting Up a Coordinate Graph

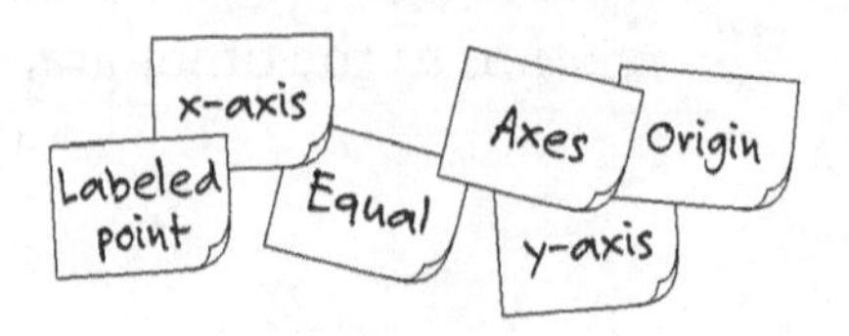

1. Use the words above to complete the sentences and to fill in the balloons. You will use some words twice.

   a. Coordinate graphs have two ______, which meet at the ______. The vertical axis is called the ______. The horizontal axis is called the _________. The scales on the axes should have ______ increments.

   b.

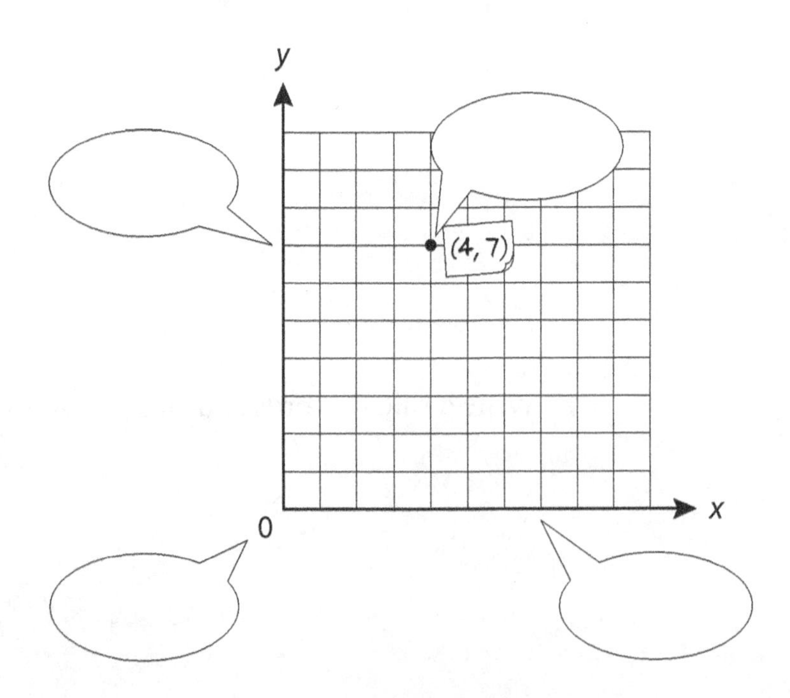

Practice: Graphing Guess My Rule

You have seen that various patterns look different when you graph them. In this activity, you will use some tables and rules such as ones you made in *Lesson 1: Guess My Rule* to make graphs and answer the questions below.

Table 1

In (x)	Out (y)
1	3
2	6
3	9
4	12

Rule 1: $y = 3x$

Table 2

In (x)	Out (y)
1	3
2	5
3	7
4	9

Rule 2: $y = 2x + 1$

Table 3

In (x)	Out (y)
1	9
2	8
3	7
4	6

Rule 3: $y = 10 - x$

1. Draw a graph for each table and its rule. Use the grid provided on page 59.
   - **a.** Label the x-axis and the y-axis.
   - **b.** Plan where you will mark the increments on each axis, and then do so.
   - **c.** Label at least three points on each graph.
   - **d.** Draw a line to connect the points.
   - **e.** Label each line with its rule or equation.
2. Describe each of the three graphs by answering the following questions:
   - **a.** What does the graph look like?
   - **b.** Which way does the line slope? Is it steep or flat?
   - **c.** Where does the graph start?
   - **d.** Describe the pattern. Does y increase or decrease as x increases?

Rule 1: Graph description

Rule 2: Graph description

Rule 3: Graph description

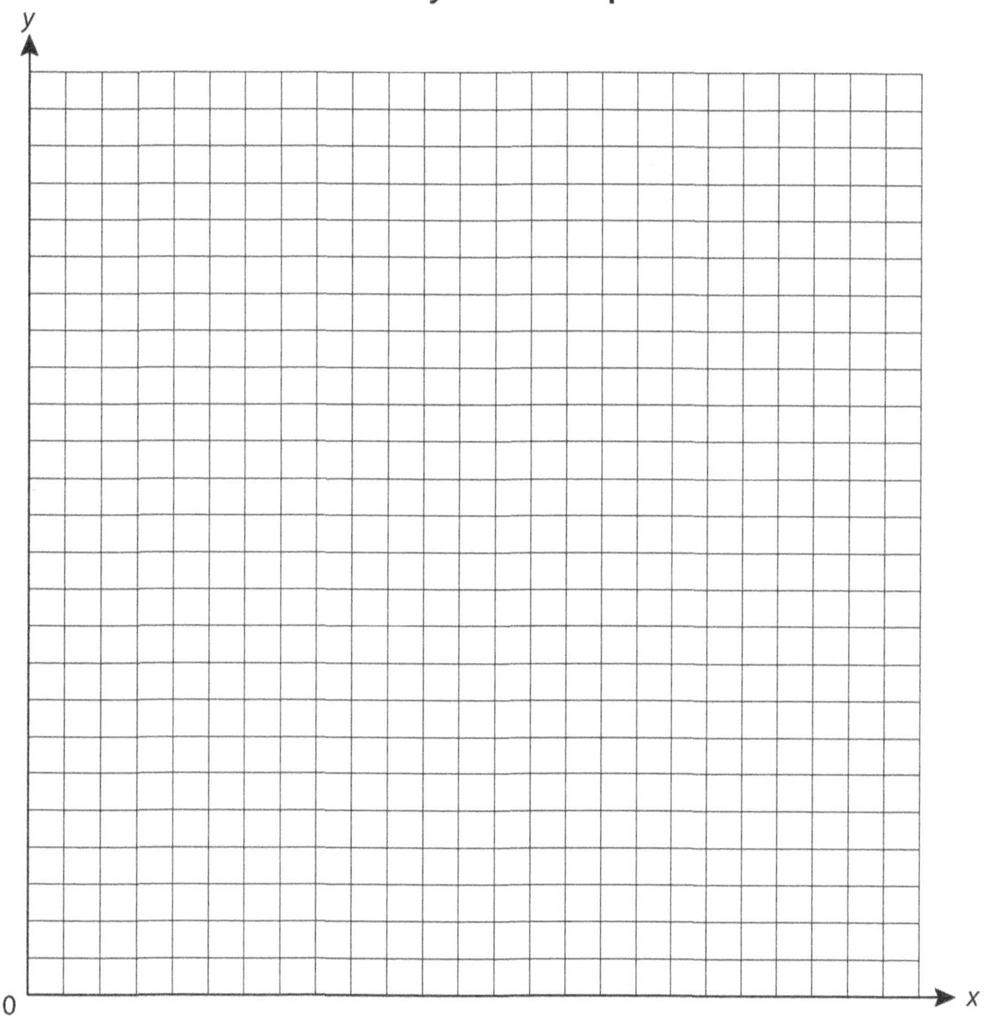

How are these graphs the same or different from the "Calories Burned" graphs on page 53?

Practice: Graphing Banquet Tables

Return to the banquet tables problem in *Lesson 2.* On the grid below, graph the equation for finding the number of people when you know the number of square tables. Remember to label the axes clearly.

People Seated at the Banquet Tables

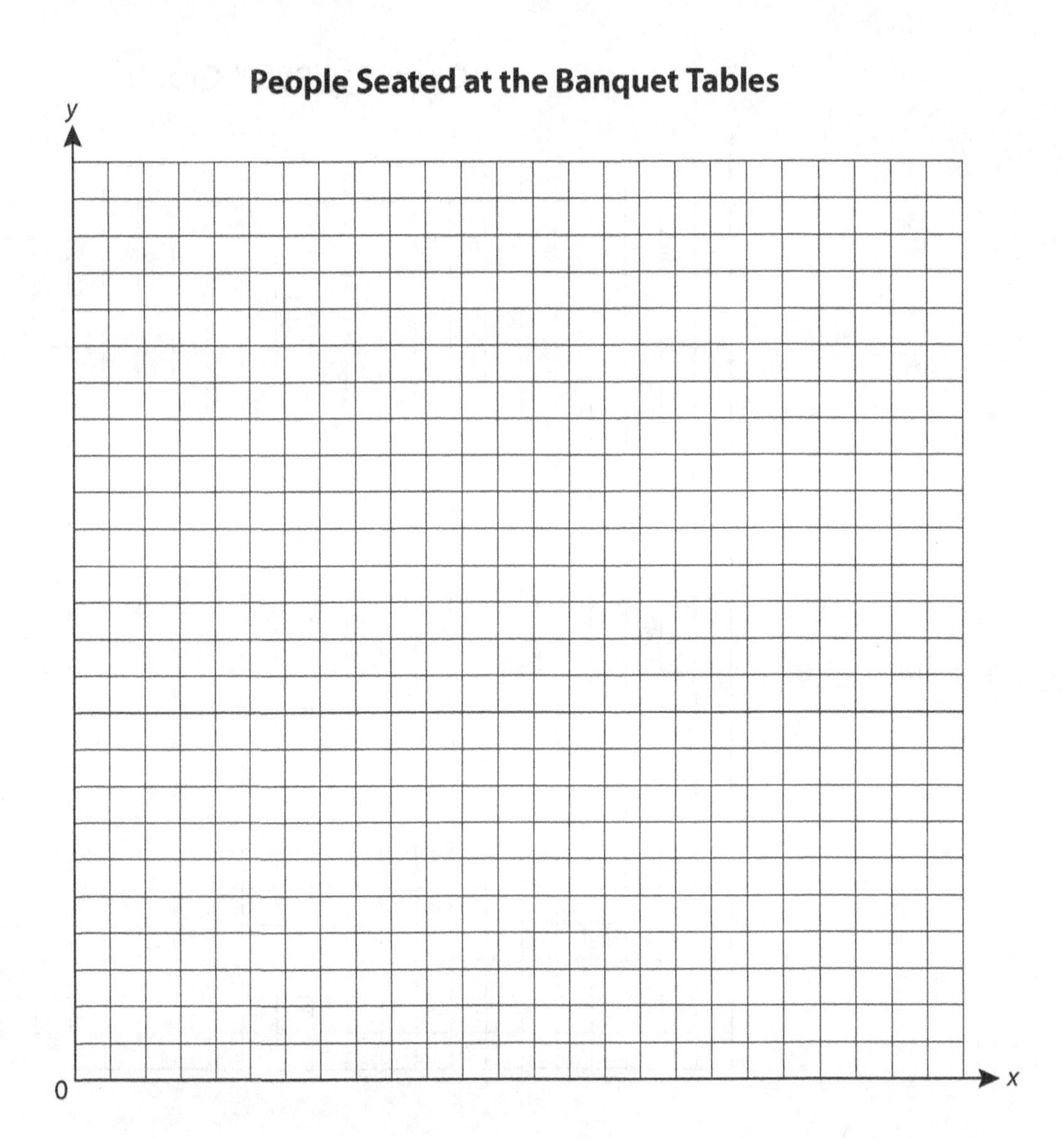

Symbol Sense Practice: Relating Multiplication and Division

Division and multiplication are called **inverse** (or opposite) **operations.** Every equation that is written with multiplication has two corresponding division equations.

5 x 4 = 20 has two corresponding division equations:

$$\frac{20}{5} = 4 \text{ and } \frac{20}{4} = 5$$

Draw a picture to show this is true.

Write the two corresponding division equations for these multiplication statements:

Multiplication Statement	Division Equation 1	Division Equation 2
1. $15 \times 20 = 300$		
2. $1{,}000 = 40(25)$		
3. $3x = y$		
4. $5m = c$		
5. $c = 20m$		
6. $m = 60h$		

What is a corresponding multiplication equation for this division statement?

7. $\frac{240}{6} = 40$

Test Practice

Questions 1 and 2 refer to the graph below.

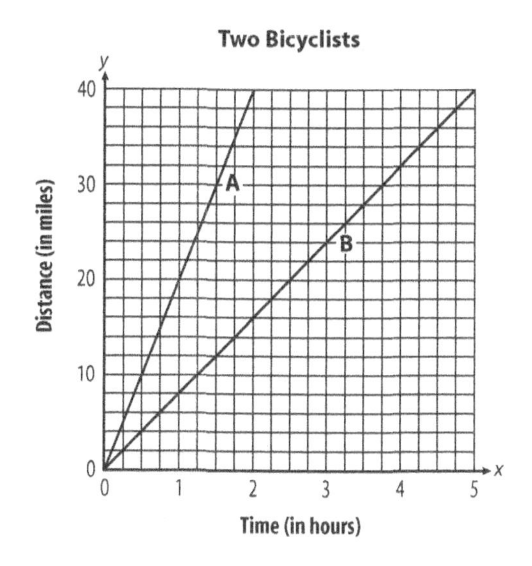

1. How many miles will bicyclist A travel in 45 minutes?

(1) 5

(2) 10

(3) 15

(4) 20

(5) 25

2. How long will it take bicyclist B to ride 10 miles?

(1) 45 minutes

(2) 1 hour

(3) 1.25 hours

(4) 1.5 hours

(5) 2 hours

3. One way to write a rule to find the distance (d) bicyclist A has traveled for any number of hours (t) is

(1) $d = 20t$

(2) $d = t + 20$

(3) $d = 15t$

(4) $d = \frac{t}{20}$

(5) $t = 20d$

Questions 4 and 5 refer to the graph below.

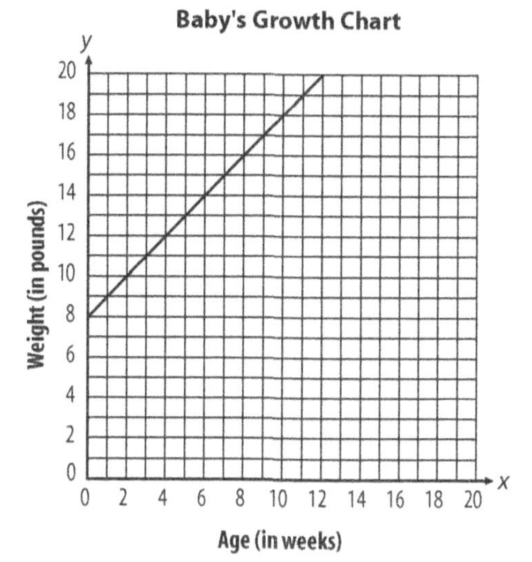

4. At what week did this baby double her birth weight?

(1) 16

(2) 8

(3) 6

(4) 5

(5) 4

5. To plot the baby's weight at nine weeks, you would mark the point

   (1) (18,8)

   (2) (17,9)

   (3) (9,17)

   (4) (9,18)

   (5) None of the above

6. Plot one point on the graph below that would follow the rule $y = 2x + 1$.

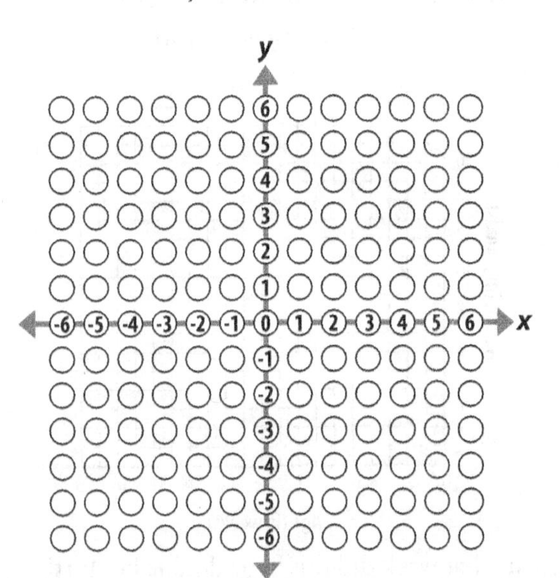

LESSON 5

Body at Work—Pushing It to the Max

What is your average maximum heart rate?

Health professionals often refer to tables to determine ideal ranges for blood pressure, heart rate, and cholesterol levels for people of different genders, ages, and weight.

In this lesson, you will use data on average maximum heart rates of people at different ages to complete a table, create a graph, and generalize a rule. You will then compare a graph and table from *Lesson 4* with the one you make today.

Activity 1: Average Maximum Heart Rates

Below is a table with average maximum heart rates listed for people of various ages.

1. Predict the average maximum heart rate for a person at age 75.
2. Predict the average maximum heart rate for a child at birth.
3. Predict the average maximum heart rate for a person your age. How did you make your prediction?
4. What patterns do you see as you look *down the columns*?
5. What patterns do you see as you look *across the rows*?
6. Use the grid paper provided to represent the table data in a graph. Label at least two points. Describe the shape of the graph.
7. How would you predict the average maximum heart rate for a person of any age? Write a rule that tells how to do this. Check your work by substituting values from the table into your rule. Does your rule work?
8. What did the researchers conclude about the relationship between age and average maximum heart rate?

Age (in years)	Average Maximum Heart Rate (in beats per minute)
20	200
25	195
30	190
35	185
40	180
45	175
50	170
55	165
60	160
65	155
70	150

Average Maximum Heart Rates

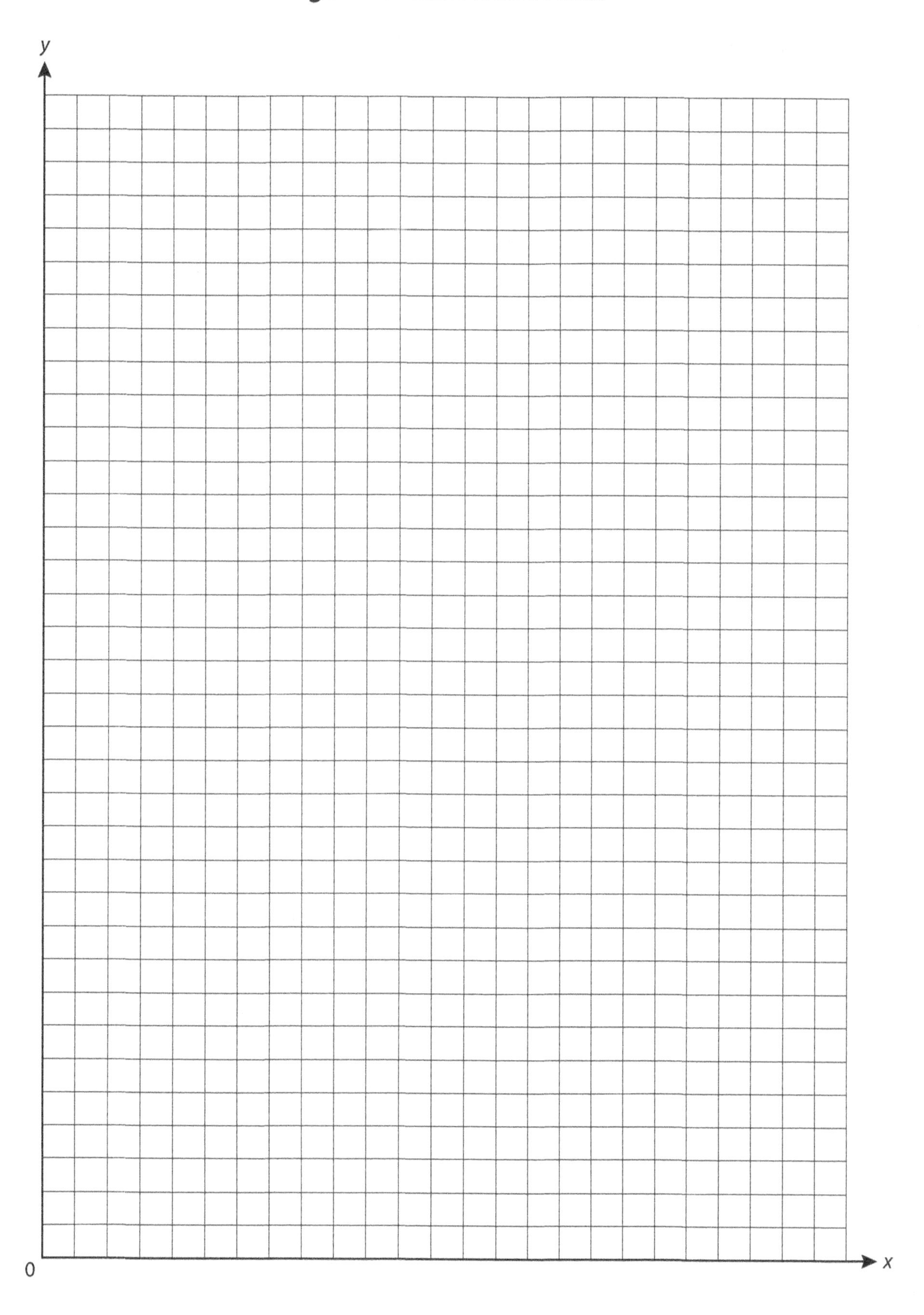

Activity 2: Comparing and Contrasting Situations

Use the "Calories Burned Running Up Stairs" table, rule, and graph from *Lessons 3* and *4* to complete this activity. Mark the pages so you can refer to them quickly.

1. Compare your *graph* of calories burned with the graph for average maximum heart rates. Look carefully at the shape of the graphs and record your thoughts below.

   a. Similarities I see:

   b. Differences I see:

2. Compare your *table* of calories burned with the "Average Maximum Heart Rates" table. Look carefully at the numbers and record your thoughts below.

   a. Similarities I see:

   b. Differences I see:

3. Compare the *equations* of calories burned with the average maximum heart rate equations. Look carefully at the symbols and numbers and record your thoughts below.

   a. Similarities I see:

   b. Differences I see:

4. In general, how do the patterns for calories burned and average maximum heart rates compare?

Practice: Brick Piles

Teon and Alicia want to build a brick walk in front of their home. They need to move all the bricks they have stockpiled in the backyard to the front-walk area before they start laying the pathway.

The table below shows their progress for part of the day.

Time of Day	Backyard Brick Pile (B)	Front-Walk Brick Pile (F)
8:00 a.m.		
10:00 a.m.	850	150
11:00 a.m.	800	200
12:00 p.m.	750	250
1:00 p.m.	700	300
4:30 p.m.		

1. Predict the number of bricks that will be in each pile at 4:30 p.m.

2. How did you make your prediction for Question 1?

3. What do you predict the graph for comparing the two piles of bricks will look like? Sketch it below.

4. On a separate piece of graph paper, draw a graph that shows the relationship between the number of bricks in the front-walk pile (F) and the number of bricks in the backyard pile (B).

5. Write two equations that describe the relationship between the two brick piles in two *different* ways.

Symbol Sense Practice: Matching Equations

Subtraction and addition are known as inverse, or opposite, operations. You see this clearly when you subtract one number from another, then check your work with addition, as shown below:

$$25 - 10 = 15, \text{ so } \ldots\ 15 + 10 = 25$$

1. Below are two columns of equations. Match each subtraction equation with its inverse operation.

Subtraction Equations	Addition Equations
1. $b - 20 = a$	**A.** $100 = 35 + 65$
2. $20 - b = a$	**B.** $a + 50 = 2b$
3. $100 - 45 = 55$	**C.** $a = 2b + 10$
4. $2b - 10 = a$	**D.** $100 = 55 + 45$
5. $2b - 50 = a$	**E.** $a + 20 = b$
6. $a - 2b = 10$	**F.** $a + 10 = 2b$
7. $100 - 65 = 35$	**G.** $a + b = 20$
8. $50 - 2b = a$	**H.** $50 = 2b + a$

2. Fill in some numbers you might substitute into the equation 2b – 10 = a to make it true.

a. If b = ___, then a = ___. **b.** If b = ___, then a = ___.

3. *Write Your Own*: Write three matching subtraction/addition equations.

Subtraction Equation	Addition Equation
a.	
b.	
c.	

Extension: Aerobic Target Heart Rate

Aerobic exercises increase your heart rate by working your muscles. Many experts believe it is good to get a 20–60 minute aerobic workout 3–5 times a week. Workouts should bring your heart rate to a range of 55%–90% of the average maximum heart rate for your age.

1. What is your heart rate at rest? ______________

2. According to the table on page 35, what is your average maximum heart rate? ______________

3. Calculate the heart rate range that experts suggest a person your age should reach when doing aerobics, based on a range of 55%–90%. Show all the work you did to get your answer.

4. Compare your calculations with the treadmill guide below. Do they agree? Explain your answer.

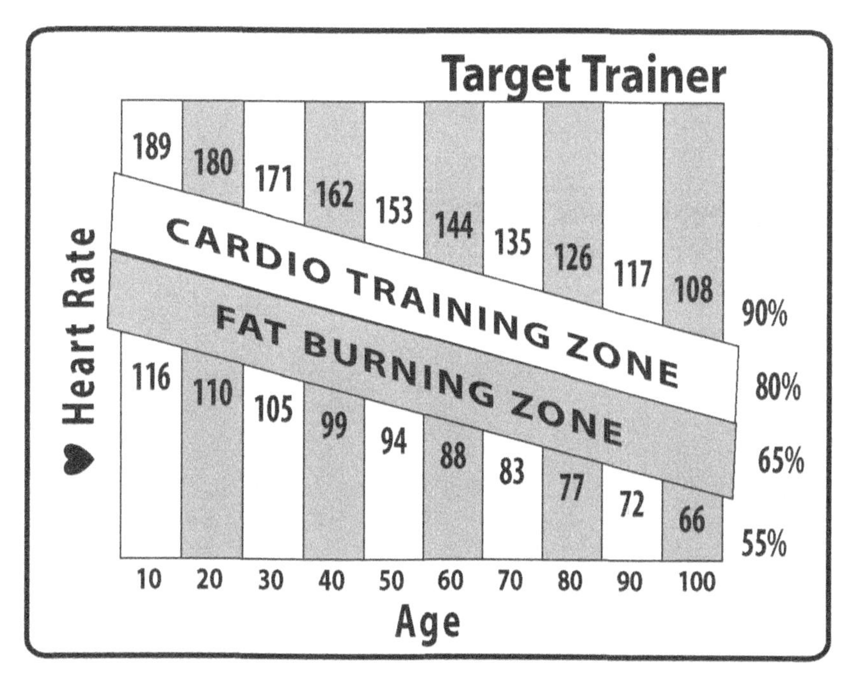

Test Practice

1. If you blink your eyes two times every five seconds, how many times will you blink your eyes in one minute?

   (1) 10

   (2) 12

   (3) 24

   (4) 60

   (5) 120

2. After spending $18 for two pizzas, Amy had $15 left. How much money did she have before she paid for the pizza? Which expression best represents this problem?

   (1) $2x + 15 = 18$

   (2) $x + 15 = 18$

   (3) $x - 18 = 15$

   (4) $18(2) = x$

   (5) $18 + x = 15$

3. A dripping faucet leaks one cup of water every half hour. How many gallons of water leak every day? (One gallon equals 16 cups.)

   (1) 1

   (2) 3

   (3) 16

   (4) 24

   (5) 48

Questions 4 and 5 refer to the information below:

Caron collected rain in a rain barrel because her state was experiencing a drought. Because it was so hot, the water in the barrel was evaporating at a rate of 1½ inches each week. The barrel started out with 36 inches of rainwater.

4. If there is no rainfall in Caron's state, how many weeks will pass before her rain barrel has 24 inches of water left? *Hint*: A table or a graph can help answer this question.

   (1) $1\frac{1}{2}$

   (2) 6

   (3) 8

   (4) 12

   (5) 24

5. Which graph shape below could most likely represent the change in the height of water in the rain barrel over the weeks of drought?

A.

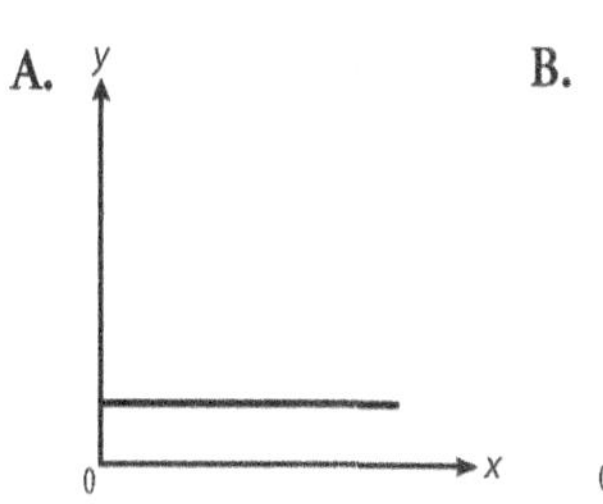

B.

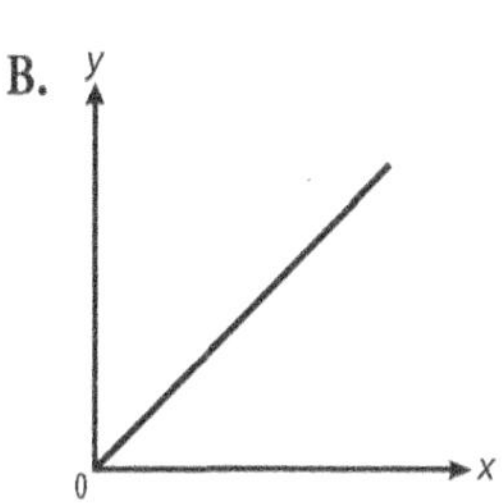

C.

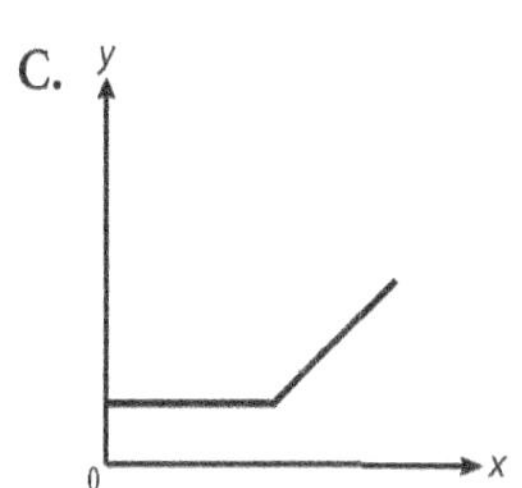

D.

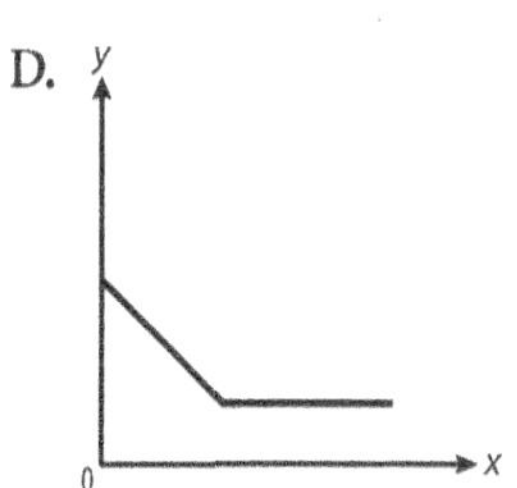

E.

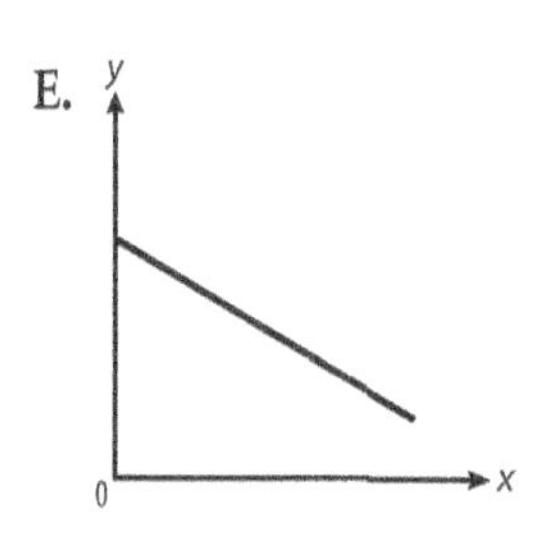

(1) A

(2) B

(3) C

(4) D

(5) E

6. On the graph below, plot the point that corresponds to the equation

$$y = x - 3, \text{ when } x = 4.$$

LESSON 6

Circle Patterns

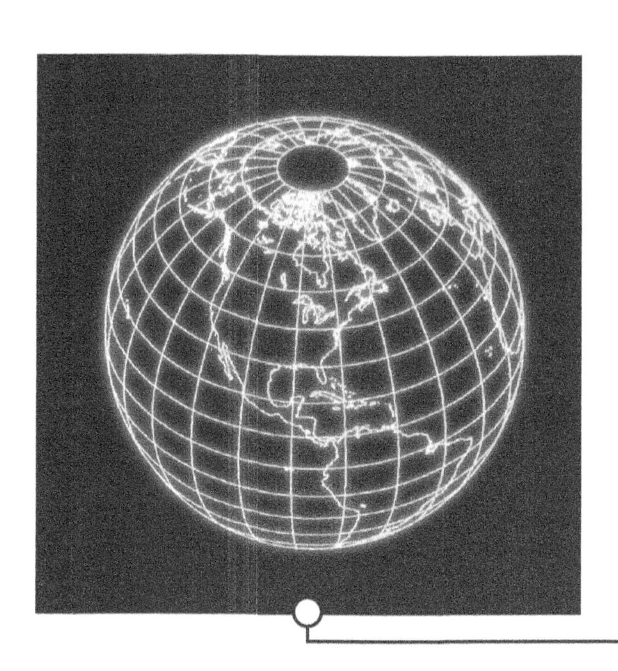

How far is it through the center of the Earth?

In this lesson, you will explore one of the most basic patterns in geometry, the circle. As you discover a **formula**, you will see how algebra and geometry relate to one another. You will be able to estimate distances around or across circular surfaces.

You will collect data by measuring circles and use it to make a table and a rule. Your rule will represent an important geometric formula.

Have you ever conducted an experiment? Today, like a scientist, you will study a situation to look for patterns and write rules that will help you solve a problem.

Think about an orange as a model of the Earth. Cut the orange in half and mentally run your finger around the outside edge. In mathematics, the line that goes around the outside edge of a circle is called the **circumference**. The imaginary line that goes around the circumference of the Earth is the equator.

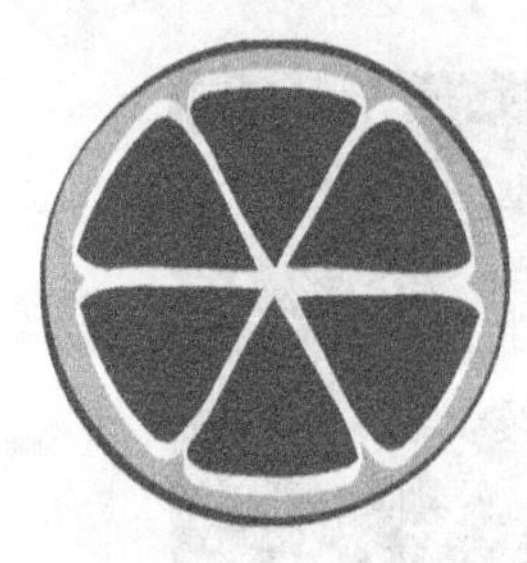

Now, imagine a line across the center of the orange, the **diameter**. How would you explain what the diameter is?

Turn the page to start your journey through the center of the Earth.

Activity 1: Journey Through the Center of the Earth

If you sail above the Earth in a balloon and travel along the equator from the Galapagos Islands, over Borneo, and then over the Congo Basin back to the Galapagos Islands, you will travel a circular route that covers approximately 23,700 miles. But if you go through the *center* of the Earth from one side of the planet to the other on the equator, how far would you travel?

Sketch the problem situation. Draw a circle to represent the Earth's equator. Draw another line for the Earth's diameter. Label the parts of your diagram with words (*circumference* and *diameter*) and any numbers you have. Write the problem question on this sketch sheet.

Investigating Circles

No one has ever traveled through the center of the Earth to measure the distance, but you can figure it out using patterns and rules about circles. To solve the mystery of the Earth's diameter, follow the steps below and answer the questions:

1. Find two circular objects.
2. Measure the circumference and diameter of each circular object using pipe cleaners or string. Cut the lengths carefully.

It will help to arrange the diameter and circumference lines one above the other like this:

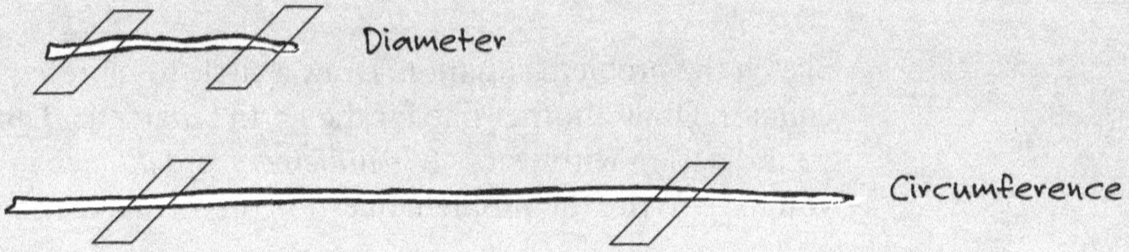

Diameter

Circumference

3. Straighten out the strings or pipe cleaners and tape them down on a paper. Label them "circumference" and "diameter." Use a ruler to measure each to the closest half-inch.
4. Make a table to keep track of circumference and diameter values. Record your data. Ask at least four other students for their data and include it in your table.

5. Describe the pattern you see in the table. Can you see that pattern in the string lengths?

6. Now write a rule that relates the circumference and the diameter.

Back to Earth

7. Use the rule relating diameter and circumference to solve the problem about the Earth's diameter. Show all work below or on your diagram of the problem. Label the length of the Earth's diameter on your diagram.

8. Add the Earth's circumference and diameter data to your table. Do the data fit the pattern of the table?

Activity 2: Measure, Calculate, and Compare

You made a table of diameter and circumference measurements. You first saw that the measured length of the circumference was a *little more than three times* the length of the diameter. The symbol for the actual number that compares the circumference length to the diameter length is π (*pi*), which is an infinite, or endless, decimal. You will often see *pi* rounded to 3.14; however, the calculator uses a slightly more exact value for *pi*.

In the table below, list four diameter and circumference measurements from the classroom data. Use your calculator to calculate the circumference for each circle given its diameter measurement.

Measured Diameter	Measured Circumference	Calculated Circumference Use $C = 3.14d$	Calculated Circumference Use $C = \pi d$
1.			
2.			
3.			
4.			

5. What do you notice when you compare the different ways to find the circumference?

Practice: Graphing Circle Data

1. Use the grid below to plot some of the data your class found from your diameter and circumference measurements in the table on page 82. Be as accurate as you can be when you plot the measurements.

   Label your x-axis "diameter measurements" and label your y-axis "circumference measurements." Pay attention to the scale you choose—you want to be able to plot all data.

2. Describe what you notice about this graph.

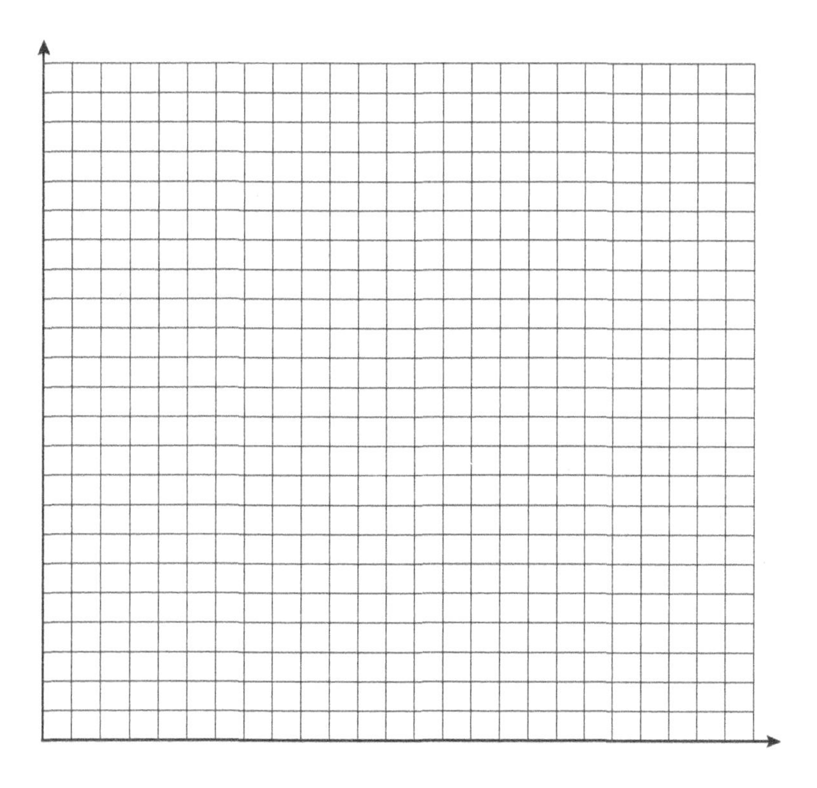

3. Plot the point on the graph with a circumference of 7 inches. About what would the diameter measure? ________

4. Plot the point on the graph with a diameter of 7 inches. About what would the circumference measure? ________

Practice: How Much More Rubber?

Bicycle tires are labeled by their diameters. How much longer is the rubber tubing for a 27-inch tire than for a 24-inch tire?

Some people say it would take 3 inches more rubber. Others think it would take about 9 inches more rubber. Who is right?

Explain your answer with a sketch.

Symbol Sense Practice: = or ≈ or < or > or ≠?

Here are five symbols that show how one quantity could be related to another:

$\approx \qquad < \qquad > \qquad \neq \qquad =$

1. Use the symbols to complete the chart below.

When you want to say…	Use this symbol
… is exactly equal to …	
… is approximately equal to …	
… is not equal to …	
… is more than …	
… is less than …	

2. Mark these statements true or false. Then rewrite all the false statements so that they become true with a correct symbol.

   a. $1 + 1 = 2$ ____________________

   b. $\$54.00 > \$44.00 + \$10.00$ ____________________

   c. $\$1.99 \times 4 \approx \$8.00$ ____________________

   d. $3 = 3.14$ ____________________

   e. $999 < 1{,}000$ ____________________

   f. $4x < x + x + x + x$ ____________________

   g. $\frac{1}{3} < \frac{1}{4}$ ____________________

   h. $\pi = 3$ ____________________

Extension: Investigating Another Pattern in Geometry

Below are pictures of five regular polygons. Examine the relationship between the number of sides of each polygon and the sum of the degrees in its angles.

Five Regular Polygons

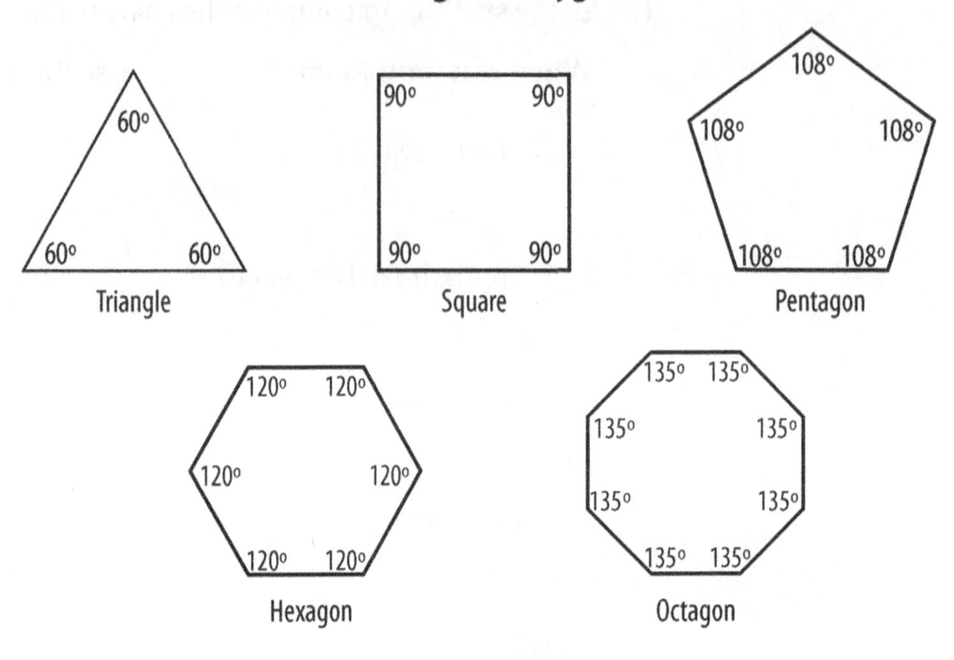

1. Record your observations in the table.

Polygon Name	Number of Sides	Sum of the Angle Measures (in degrees)
triangle		
square		
pentagon		
hexagon		
octagon		

2. Which of these is the rule for finding the sum of the angles of any regular polygon(S) when you know the number of sides (n)?

   a. $S = 60n$

   b. $S = 180\ (n - 2)$

   c. $S = 90\ (n - 2)$

   d. $S = n\ (n - 2)$

3. Show that the rule you chose works for three of the shapes.

Test Practice

1. A tree has a 2-foot diameter. About how long would a string be if you wrapped it once around the tree?

   (1) 8 inches

   (2) 2 feet

   (3) 4 feet

   (4) 6 feet

   (5) 8 feet

2. About how much longer is the circumference of a $2\frac{1}{2}$ foot diameter tree than the circumference of a 2-foot diameter tree?

   (1) About $\frac{1}{2}$ foot

   (2) About $1\frac{1}{2}$ feet

   (3) About 3 feet

   (4) About 5 feet

   (5) About $7\frac{1}{2}$ feet

3. According to the table below, which rule below shows how to find the number of hours driven for any number of miles? Let t stand for the time and d stand for distance driven.

Time Driven	Distance Driven
1 hr.	
2 hr.	
3 hr.	135 mi.
4 hr.	
5 hr.	
6 hr.	270 mi.

   (1) $t = d - t$

   (2) $t = 45d$

   (3) $t = d + 45$

   (4) $t = \frac{d}{45}$

   (5) $t = 135d$

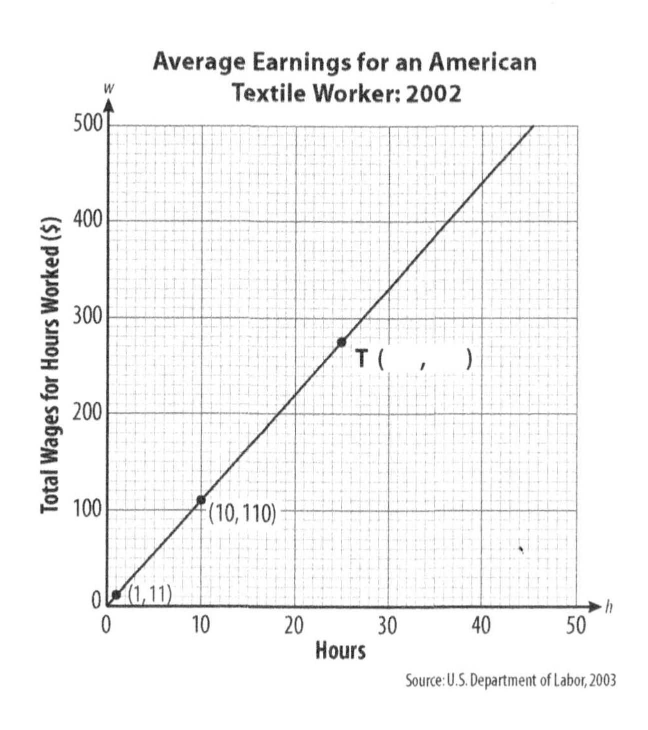

Refer to the graph above to answer Questions 4 and 5.

4. The correct label for point *T* in the graph is

(1) (275, 25)

(2) (25, 275)

(3) (23, 203)

(4) (10, 140)

(5) (1, 14)

5. If *h* stands for hours worked and *w* stands for hourly wage, which equation below represents the same pattern as the graph?

(1) $40w = h$

(2) $11w = h$

(3) $40h = w$

(4) $w = 11h$

(5) $w = 11 + h$

6. 3.14 x 10 =

LESSON 7

What Is the Message?

What are they trying to tell me?

Behind every equation is a message that could describe a pattern in a real-life situation. Seeing that message makes the equation come alive. An equation also provides clues about what a graph could look like and what data you could expect to find in a table created from it.

In this lesson you will practice translating between equations and verbal rules. You will also look at two kinds of equations and compare their patterns, so you can better understand the information equations communicate.

Activity 1: Pass the Message

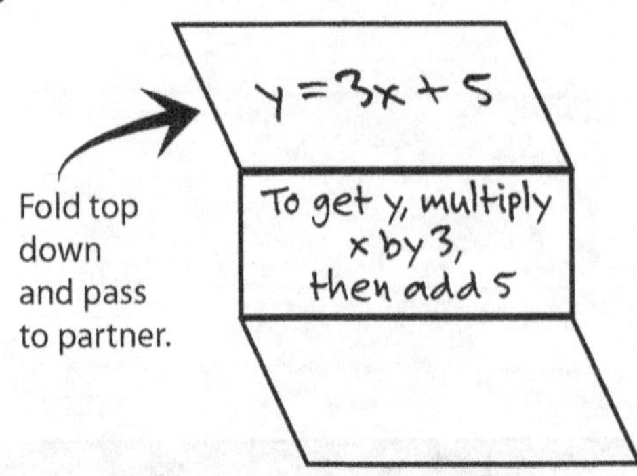

You will work on this activity with a partner, passing your work back and forth.

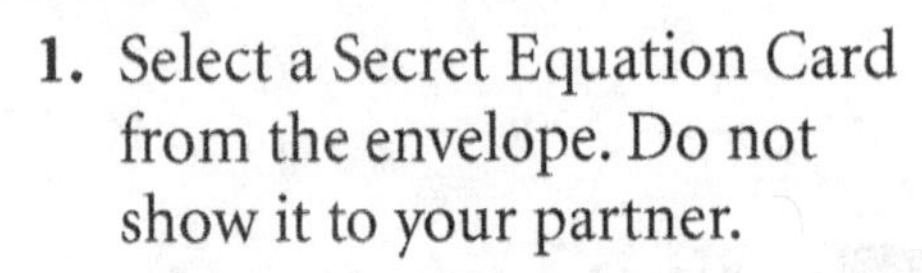

1. Select a Secret Equation Card from the envelope. Do not show it to your partner.
2. Write the equation on the top section of the paper provided.
3. On the middle section of the paper, translate the equation into a rule in words. Use a complete sentence.
4. Fold the top section of the paper under so the equation does not show, and exchange papers with your partner.
5. Read your partner's sentence without looking at the equation, and then write an equation that matches the sentence on the bottom section of the paper.
6. Repeat these steps for three more equations.

Examine the four papers with your partner, and then answer the questions for each equation:

Did the equations match? Why/why not?

Top Equation	Bottom Equation	Where It Did or Did Not Match

Activity 2: What Is the Message?

All kinds of equations and formulas are used by people every day. Every equation sends a message. In this activity, you will think about the messages behind two equations.

With a partner, select one pair of equations.

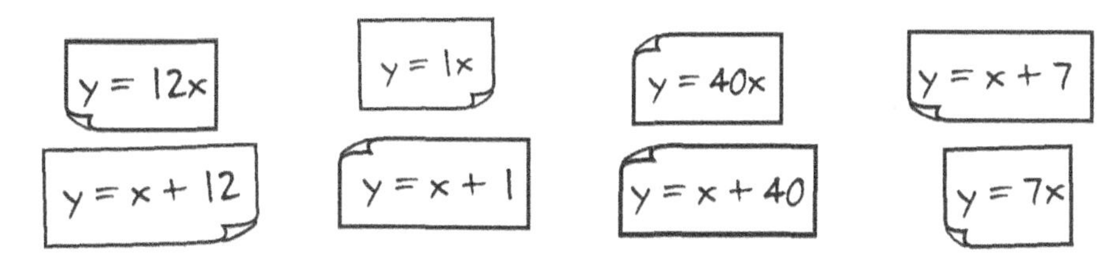

Illustrate each equation with a table and graph. Draw and label the graphs on the same set of axes. Do your multiplication equation first.

Make the math come alive! Tell a brief story or describe a situation that might correspond to each equation and its table and graph. You can choose to demonstrate with a diagram or objects instead of writing.

> Remember, these numbers and variables (x and y) could stand for anything—for example, money, miles, ages, inches, or objects.

Prepare a poster of your results to share with the class.

Your multiplication equation: ____________________

1. The table of your equation:

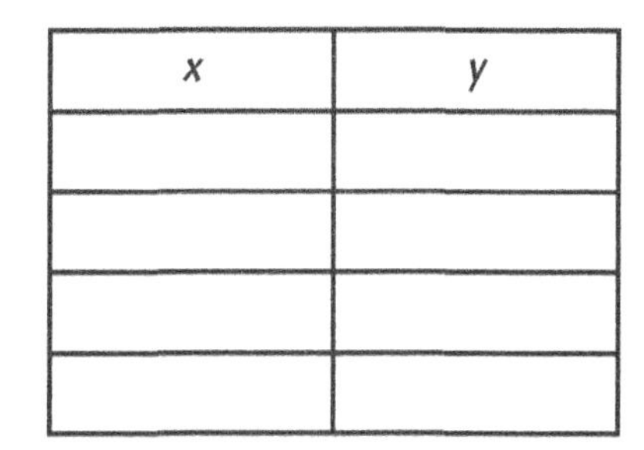

x	y

2. The graph of your equation. Use a separate piece of graph paper.

3. The description of the situation your equation, table, and graph could represent. Be clear about what x and y stand for.

4. The diagram of your equation situation:

Your addition equation: ______________________

1. The table of your equation:

x	y

2. The graph of your equation. Use the same piece of graph paper you used for the multiplication equation.

3. The description of the situation your equation, table, and graph could represent. Be clear about what x and y stand for.

4. The diagram of your equation situation:

Practice: Geometry Formulas

A **formula** is a standard rule used to find a specific quantity—for example, the area of a square or a rectangle, the circumference of a circle, or the volume of a cube. Below you will find a boxed key that gives the meaning of several variables used in geometry. Use these definitions to translate the equations from symbols into words. One example is provided.

Key to Variables Used in Geometry

l = length w = width h = height s = side
P = perimeter A = area V = volume
C = circumference d = diameter r = radius

For Rectangles and Rectangular Solids:

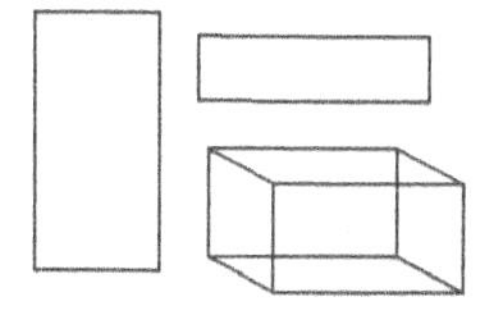

$P = 2l + 2w$ ________________________________

$A = lw$ The area of a rectangle equals the length times the width.

$V = lwh$ ________________________________

For Squares:

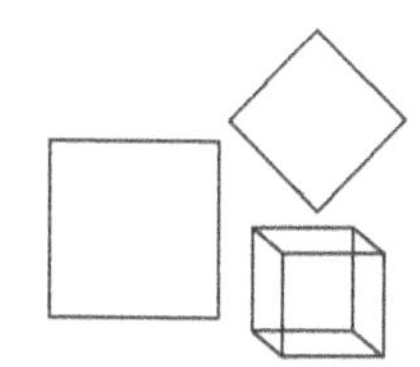

$P = 4s$ ________________________________

$A = s^2 = s \times s$ ________________________________

$V = s^3 = s \times s \times s$ ________________________________

For Circles:

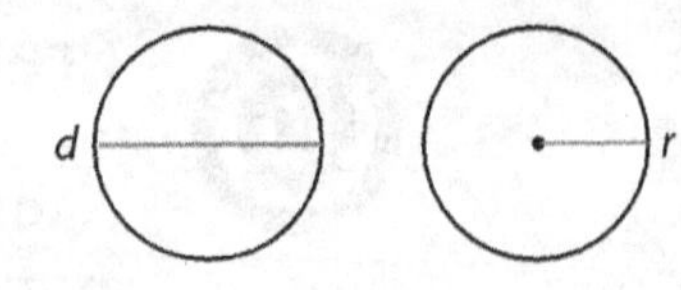

$C = 2\pi r$ ____

$C = \pi d$ ____

$d = \frac{C}{\pi}$ ____

$A = \pi r^2 = \pi \times r \times r$ ____

Define the following terms with pictures and words:

Perimeter:

Area:

Volume:

Circumference:

Diameter:

Radius:

Symbol Sense Practice: Evaluating Geometric Formulas

Find the value using the geometric formula and values provided. Show your work.

1. $A = lw$ Find A when $l = 40$ and $w = 60$.	**2.** $P = 2l + 2w$ Find P when $l = 90$ and $w = 10$.	**3.** $A = s^2 = s \times s$ Find A when $s = 10$.
4. $P = 4s$ Find P when $s = 10$.	**5.** $V = lwh$ Find V when $l = 20$, $w = 18$, and $h = 5$.	**6.** $V = s^3 = s \times s \times s$ Find V when $s = 10$.
7. $P = 2(l + w)$ Find P when $l = 100$ and $w = 62$.	**8.** $C = \pi d$ Find C when $d = 10$. (Use $\pi = 3.14$)	**9.** $A = \pi r^2 = \pi \times r \times r$ Find A when $r = 10$. (Use $\pi = 3.14$)
10. $C = 2\pi r$ Find C when $r = 10$. (Use $\pi = 3.14$)	**11.** $A = lw$ Find A when $l = 3.5$ and $w = 6$.	**12.** $P = 2l + 2w$ Find P when $l = 12.5$ and $w = 7.5$.

Test Practice

1. Which situation below matches the equation $y = 24x$?
   (1) The total number of cases (y), for any number of soda cans (x) packed 24 to a box
   (2) The total number of cans of soda (y), packed 24 cans to a carton, in any number of cartons (x)
   (3) The total number of months (x) in any number of years (y)
   (4) The total gas used by a car that travels 24 miles
   (5) The total number of \$24 shirts (x) that could be bought with y dollars

2. Which graph below could represent the equation $y = 10x$?

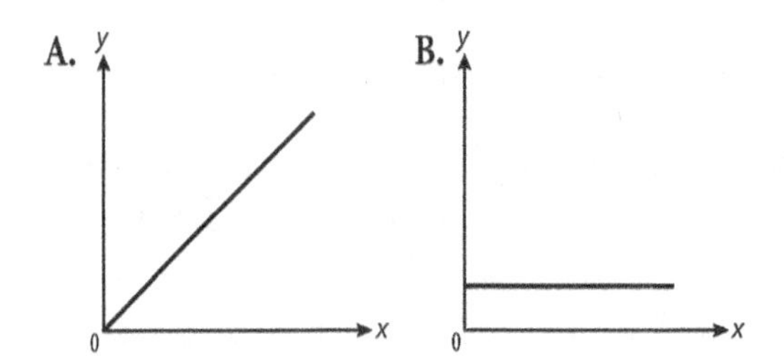

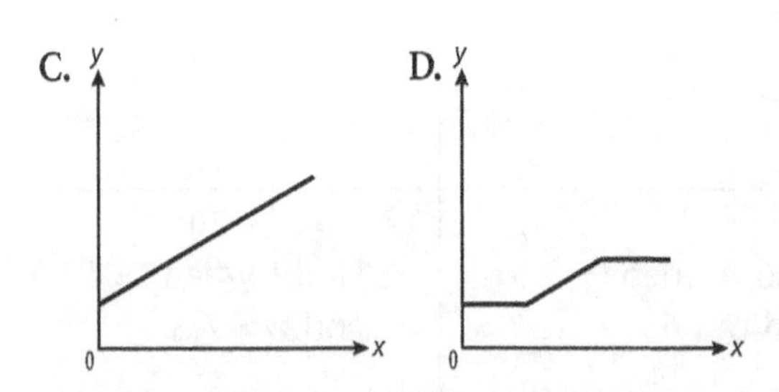

   (1) Graph A
   (2) Graph B
   (3) Graph C
   (4) Graph D
   (5) None of the above

3. Which equation below could be used to describe the total number of inches (y) in any number of yards (x)?
   (1) $y = \frac{x}{36}$
   (2) $y = 36 - x$
   (3) $y = 36x$
   (4) $y = 36 + x$
   (5) $y = \frac{x}{12}$

Questions 4 and 5 refer to the information below:

The rule of thumb sometimes used to estimate temperatures when converting from Celsius to Fahrenheit degrees is "double the Celsius temperature, and add 30°."

4. Which table below follows the rule stated above?

A.

Celsius	Fahrenheit
15	60
32	94
50	130

B.

Celsius	Fahrenheit
60	15
94	32
130	50

C.

Celsius	Fahrenheit
15	45
32	64
50	80

D.

Celsius	Fahrenheit
15	0
32	34
50	70

(1) Table A

(2) Table B

(3) Table C

(4) Table D

(5) None of the above

5. Which equation represents the rule as stated above?

(1) $F = 2(C + 30)$

(2) $F = 2C + 30$

(3) $C = 2(F + 30)$

(4) $C = 2F + 30$

(5) $F = \frac{C}{2} + 30$

6. A more exact formula for changing from Celsius to Fahrenheit degrees is $F = 1.8C + 32$. What would the Fahrenheit equivalent of 10°C be?

LESSON **8**

Job Offers

How can I compare these job offers?

Have you ever had to make a hard financial decision by considering how something will play out in the long run? It is helpful to be aware of ways to look at and compare different options. Algebra offers some useful tools for analyzing situations over time in order to make decisions.

In this lesson, you will solve a problem using arithmetic and then make a table and a graph about job offers. You will compare two situations to make an informed decision about which job is better.

Activity: Job Offers

Who is right?

What a great week for Armand! He was offered both part-time jobs for which he had applied. Now he needs to decide which one to take. He told his partner, Cheri, that LaserLink offered to pay him $200/week whereas QuinStar's offer was to pay $150/week plus a $2,000 sign-on bonus that he would get before he even started working.

> "I am going to take the QuinStar job," he said, "because I do not intend to have to hold a second job for more than a year."
>
> Cheri replied, "I don't know, Armand. I think you are being foolish. In a year, you would make more money at LaserLink. I bet you would make more money at LaserLink in six months!"
>
> "No way! I will prove it to you." Armand shouted.

Who is right, Armand or Cheri?

Show how you know.

Would They Ever Be the Same?

The Situation: Armand would start out making more money at QuinStar; but by the end of the year, he would have made more money working for LaserLink.

1. **The Question:** Is there ever a point in time when the accumulated earnings from each job would be the same?

Make a table and a graph to solve the problem. Start with whichever one you choose, but make both. Below are some suggestions for how to show the accumulated pay for each job. Use separate pieces of paper for the table and the graph.

Table	Graph
Make a table showing weeks worked and pay accumulated for each job. Remember to account for the sign-on bonus that QuinStar would pay before the first week's pay. *Note:* You should have entries that cover up to 52 weeks, but you do not need an entry for every week. You may also make a three-column table with entries for both jobs side-by-side for easy comparison.	Make a graph. Use increments of two weeks across the *x*-axis (2, 4, 6 weeks, etc.), and use increments of $500 on the *y*-axis. Plot a few points for each job. Use different colors for the two jobs. For each job, draw the line that represents the relationship between that job's pay and the weeks worked.

2. Write a rule in words and/or in symbols to show how much money Armand would make at LaserLink for any number of weeks he worked.

3. Write a rule in words and/or in symbols to show how much money Armand would make at QuinStar for any number of weeks he worked.

4. Describe what you notice by answering these questions about the situations:

   a. What stays the same?

   b. What changes?

   c. What stands out for you?

Practice: The Race

Calvin and Jackie plan a race through town. Calvin is the slower runner, so Jackie offers him a two-mile head start. If Calvin runs at 3 mph and Jackie runs at 4 mph, when will Jackie catch up with Calvin during the 12-mile course?

Use a table and a graph to support your answer.

> Where graph lines intersect is important because it shows the point at which two situations would have the same outcome.
>
> Where a graph starts is also important. Pay attention to where the starting point shows up in the equation.

Practice: Armand's Weeks, Not Pay

The rule for finding Armand's pay for any number of weeks he worked at QuinStar is $p = 150w + 2000$	The rule for finding Armand's pay for any number of weeks he worked at LaserLink is $p = 200w$

If you knew the amount of money Armand had made, but you did not know how long he had worked, how could you figure out the number of weeks he had been at either one of the jobs?

1. Armand has earned $14,000 working at LaserLink. How many weeks has he worked? Show all your work.

2. Armand has earned $10,800 at LaserLink. How many weeks has he worked? Show all your work.

3. Write an equation that shows how to find the number of weeks worked for any amount of pay Armand earned at LaserLink.

4. Armand has earned $17,000 at QuinStar. How many weeks has he worked? Show all your work.

5. Armand has earned $10,400 at QuinStar. How many weeks has he worked? Show all your work.

6. Write an equation that shows how to find the number of weeks worked for any amount of pay Armand earned at QuinStar.

Symbol Sense Practice: Solving One-Step Equations

An equal sign tells you that the quantities on either side of it are equal; the quantities balance. When you solve an equation, you try to see what number will make the sides of the equation balance.

What must x be to balance these equations? For Questions 1–10, find the missing number in the equation.

1. $x + 13 = 30$ $x =$ _____ **2.** $1{,}100 = x + 90$ $x =$ _____

3. $x - 50 = 150$ $x =$ _____ **4.** $75 = x - 75$ $x =$ _____

5. $4x = 100$ $x =$ _____ **6.** $10x = 1{,}000$ $x =$ _____

7. $600 = 100x$ $x =$ _____ **8.** $x + \frac{1}{2} = 5$ $x =$ _____

9. $\frac{x}{3} = 7$ $x =$ _____ **10.** $\frac{x}{10} = 88$ $x =$ _____

Balance the amounts. For Questions 11–20, write an equation for each statement. Then find the missing number that balances the equation.

11. If you add 7 to a certain number, you will get 20.

The equation is _____________________. $x =$ _____.

12. If you add a certain number to itself, you will get 100.

The equation is _____________________. $x =$ _____.

13. If you add 100 to a certain number, you will get 6,875.

The equation is _____________________. $x =$ _____.

14. If you subtract a certain number from itself, you will get 0.

The equation is _____________________. $x =$ ____.

15. If you multiply a certain number by 8, you will get 56.

The equation is ______________________. $x =$ ______.

16. If you multiply a certain number by itself, you will get 100.

The equation is ______________________. $x =$ ______.

17. If you double a certain number, you will get 1,200.

The equation is ______________________. $x =$ ______.

18. If you subtract a certain number from 27, you will get 10.

The equation is ______________________. $x =$ ______.

19. If you divide a certain number by 2, you will get 75.

The equation is ______________________. $x =$ ______.

20. If you divide a certain number by 10, you will get 100.

The equation is ______________________. $x =$ ______.

Extension: What If?

Changes in a situation result in changes in a graph of that situation. Below, you are asked to consider some changes to the job offer situation explored in this lesson. Think about the graph and what would happen if…

1. **What if QuinStar had offered a $3,000 bonus instead of a $2,000 bonus? What would the graph for QuinStar and LaserLink look like?**

   Sketch your work on a separate piece of graph paper and include an explanation of your sketch. You may also make tables that reflect the new circumstances and develop a graph from the table data.

2. **What if neither LaserLink nor QuinStar had offered a bonus? What would the graph of both job offers look like then?**

   Sketch your work on a separate piece of graph paper and include an explanation of your sketch. You may also make tables that reflect the new circumstances and develop a graph from the table data.

3. **What if LaserLink had offered the bonus, instead of QuinStar? What would the graph of both job offers look like then?**

   Sketch your work on a separate piece of graph paper and include an explanation of your sketch. You may also make tables that reflect the new circumstances and develop a graph from the table data.

Test Practice

1. A person earns $10 an hour at a full-time job and earns a $100 bonus for being employee of the week. Which expression below represents how much that person will earn in a month with four paychecks? (Assume the person works 40 hours per week.)

   (1) 10(4) + 100

   (2) 10(40)(4)

   (3) 10(24)(4) + 100

   (4) 10(40)(4) + 100

   (5) 10(40/4)

2. Which graph below could show the growth of two plants if one of the plants was started from seed and the other was started from a 2-inch cutting?

A.
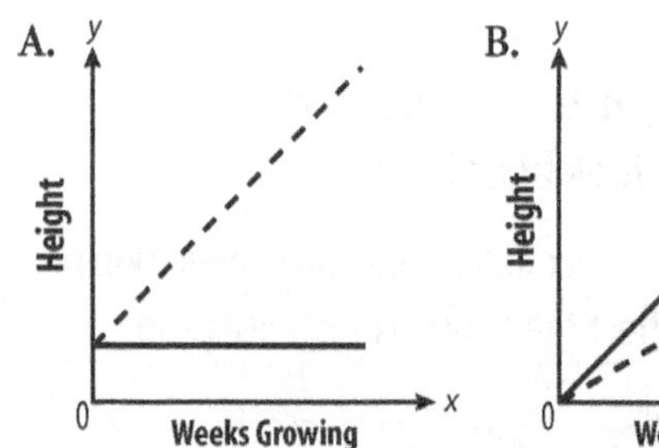

B.
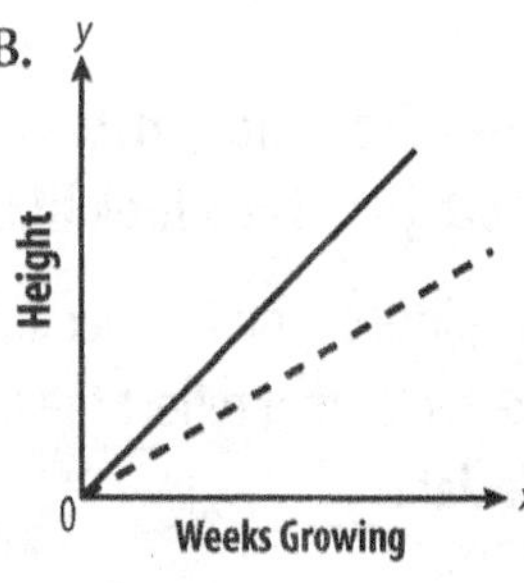

C.
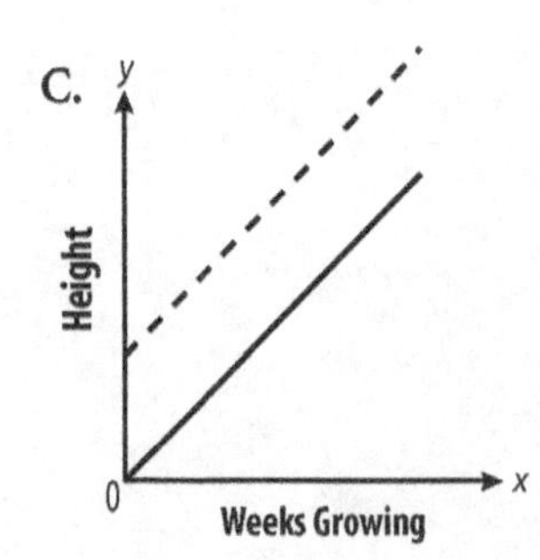

D.
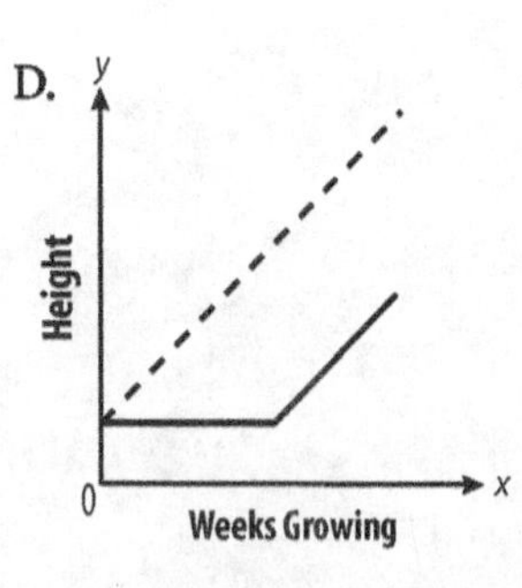

   (1) A

   (2) B

   (3) C

   (4) D

   (5) None of the above

3. The chart below shows the gas mileage for a sport utility vehicle (SUV). Notice the relationship between distance driven and gas used during the trip.

Miles Driven	Gallons Used
50	12
240	48
300	60

   Which rule shows the total miles (M) this SUV will get for any number of gallons of gas used? Let G equal the number of gallons used.

   (1) $M = 12 + G$

   (2) $M = \frac{G}{5}$

   (3) $M = 5(G)$

   (4) $M = G + 48$

   (5) $M = 60G$

4. Amos turned on the faucet to fill the large water tub in his lab, so he could perform an experiment on waves. He skipped two entries. What are the missing entries?

Time (minutes)	3	6	12	15	21
Water Level (inches)		4	8		14

   (1) 15 and 2

   (2) 5 and 17

   (3) 13 and 1

   (4) 2 and 10

   (5) 2 and 12

5. Sharyn buys three cans of beans. Each can costs \$1.29. She has a coupon for \$0.50 off her total cost. Which expression below shows what Sharyn will pay for the beans?

(1) 3 + \$1.29 – \$0.50

(2) 3 + \$1.29 + \$0.50

(3) \$1.29 – \$0.50

(4) \$1.29 – 3(\$0.50)

(5) 3(\$1.29) – \$0.50

6. A map of the United States is drawn to scale such that a 3″ line represents 450 miles. How many miles does a 1″ line represent?

LESSON **9**

Phone Plans

How do I know which plan is best for me?

Ads are everywhere—luring us to choose a long-distance phone plan or to take a loan on credit, for example. It is not always easy to figure out which company offers the best deal. Your algebra tools will help you see more clearly which deal is best.

You will piece together information about four phone plans. To find the best deal, you will have to make sense of tables, graphs, words, and equations. Then you will think about the advantages and disadvantages of the plans for particular customers.

Activity 1: Phone Plans

Four ads for cellular and long-distance phone plans came in the mail today, but the ads got ripped up. Can you put the pieces back together? Your teacher will give you all of the words, tables, and equations from the ads. Reattach them to their matching graphs.

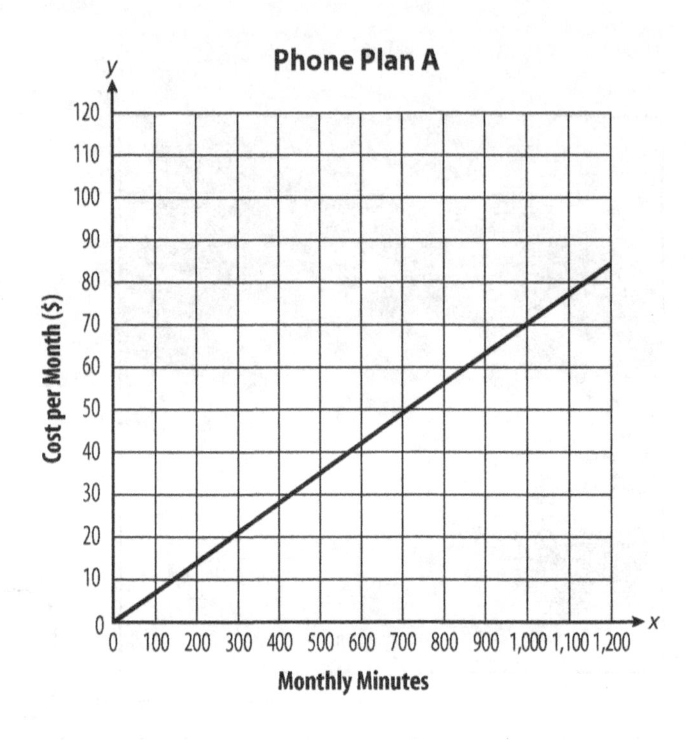

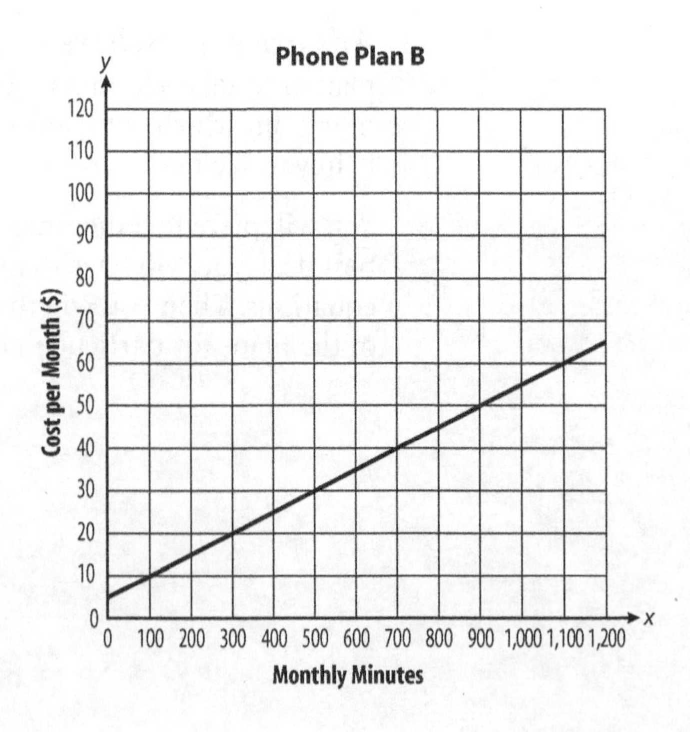

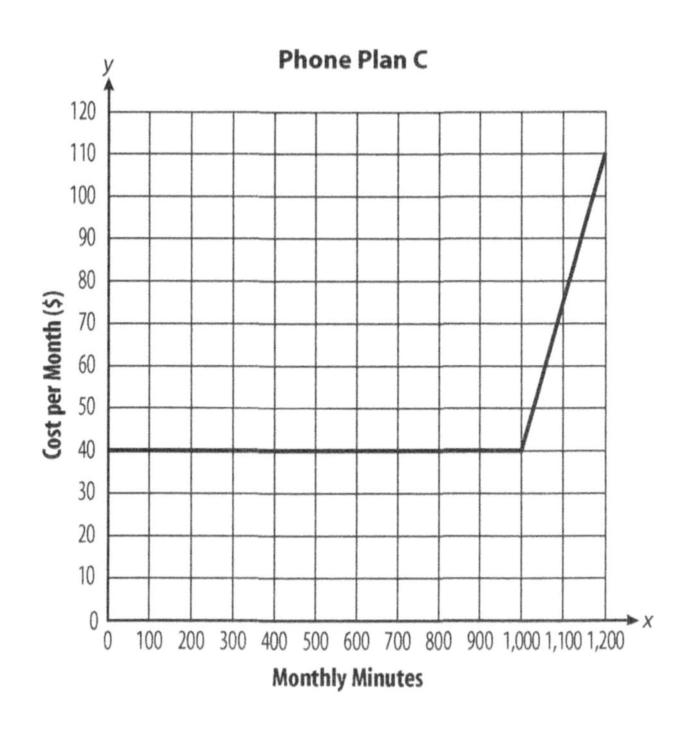
Phone Plan C
y
120
110
100
90
80
70
60
50
40
30
20
10
0
Cost per Month ($)
0 100 200 300 400 500 600 700 800 900 1,000 1,100 1,200
x
Monthly Minutes

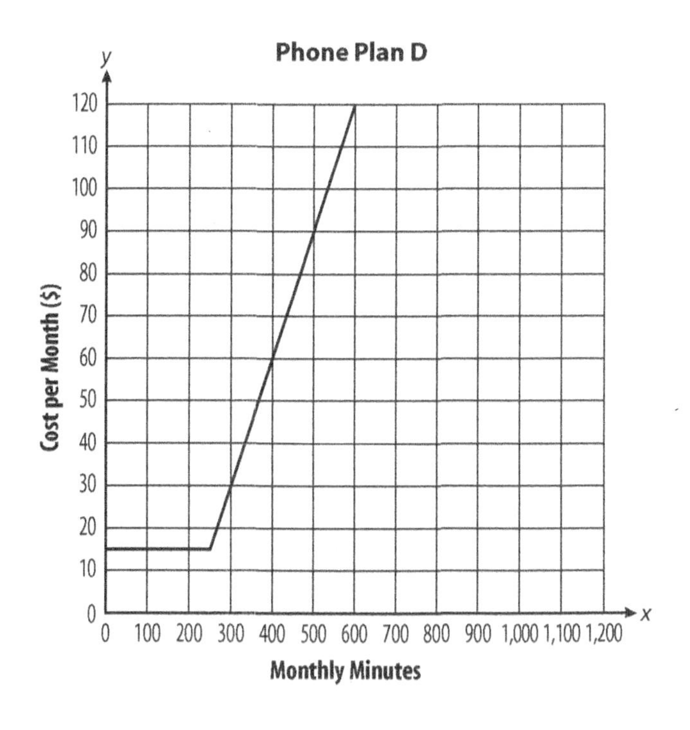
Phone Plan D
y
120
110
100
90
80
70
60
50
40
30
20
10
0
Cost per Month ($)
0 100 200 300 400 500 600 700 800 900 1,000 1,100 1,200
x
Monthly Minutes

Activity 2: It Would Depend on the Person

Use the pieced-together advertisements to answer the following questions about which plans are best and worst for specific customers:

1. Mary Jane loves to talk to her family, many of whom live out of state. She wants as many minutes as she can get, but definitely does not want to pay more than $50.00. Which plan should she choose? Why? Which plan would be the worst for her? Why?

2. Jenny, who lives in New York City, talks to her best friend in Miami every night for about a half-hour. Other than that, she makes very few long distance calls. Which plan is best for her and which is the worst? Why?

3. Tricia wants to get a phone she will only use in emergencies, for instance, if her car breaks down. Which plan is best for her and which is the worst? Why?

4. Which phone plan would be the best for you? Which would be the worst for you? Why?

Practice: I Am Changing the Rules!

You are the new Chief Executive Officer (CEO) of Cellular Wireless that has advertised Plan A, and you think it is time to change the rules.

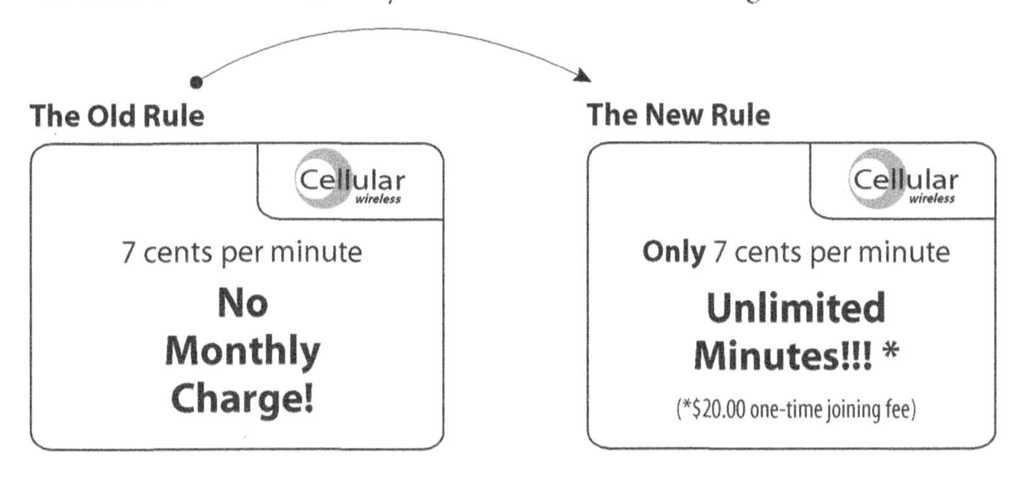

Change the old equation, table, and graph to go along with your new rule for the first month.

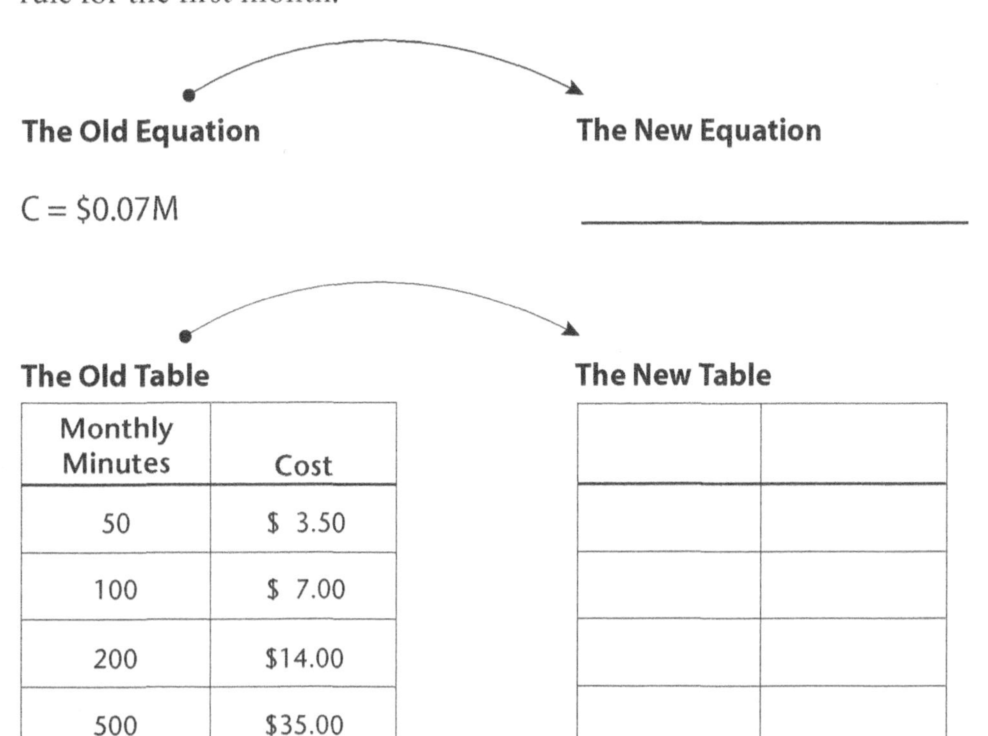

The Old Equation

$C = \$0.07M$

The New Equation

____

The Old Table

Monthly Minutes	Cost
50	$ 3.50
100	$ 7.00
200	$14.00
500	$35.00

The New Table

This is a graph of the old plan.

Graph the new plan here in another color.

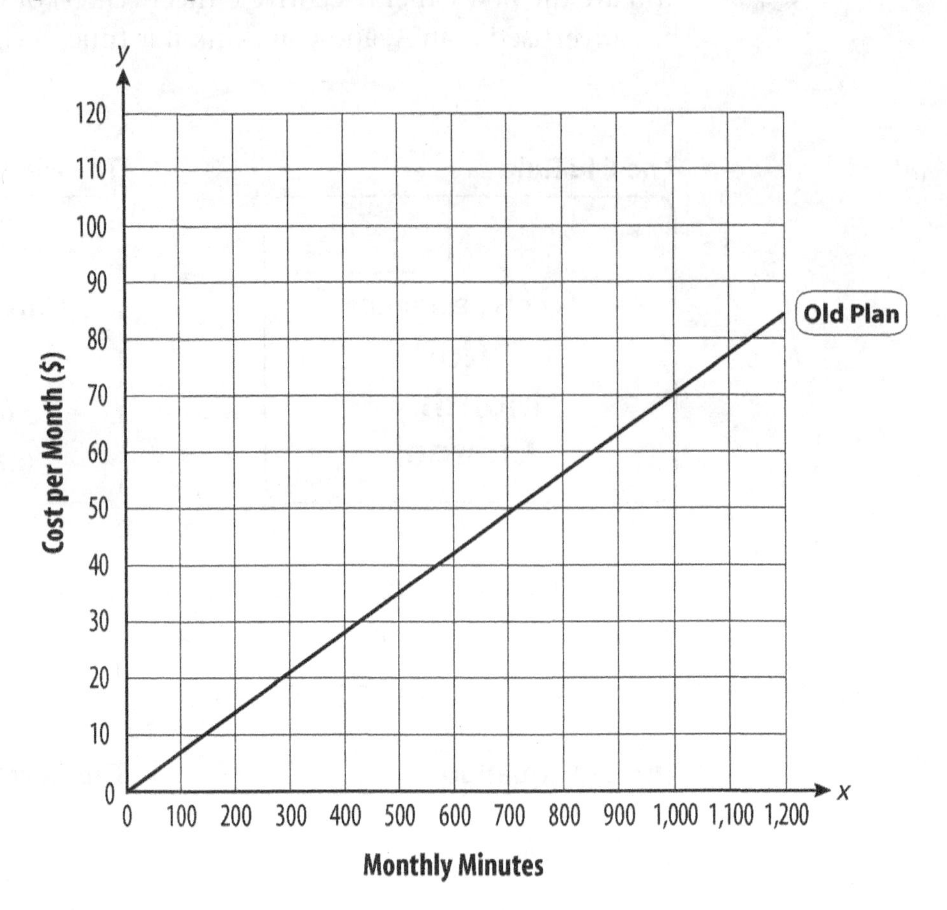

Symbol Sense Practice: Greater Than, Less Than

Math Symbol	In Words
$=$	equals
$>$	is greater than
$<$	is less than
$\geq$	is greater than or equal to
$\leq$	is less than or equal to

Use the above math symbols to rewrite the advertisements below where cost depends upon age. Let C stand for cost, and let A stand for age.

Advertisements / **Math Symbols**

1. Movie tickets

Under 12 years: \$5.00 — When $A < 12, C = \$5.00$

12 and over: \$9.00 — When $A \geq 12, C = \$9.00$

2. Coffee at a diner

Cup of coffee: \$1.25

Senior citizens (65 and over): \$0.05

3. Club membership

Under 21: free

21 and over: \$10.00

4. Airline tickets

Five and over: \$250

Under five years: half price

Symbol Sense Practice: Solving Two-Step Equations

One easy way to solve two-step equations is to cover part of the equation using the "finger cover-up" method.

Faced with an equation such as $2x + 3 = 19$, put your finger over the "$2x$" part of the equation like this:

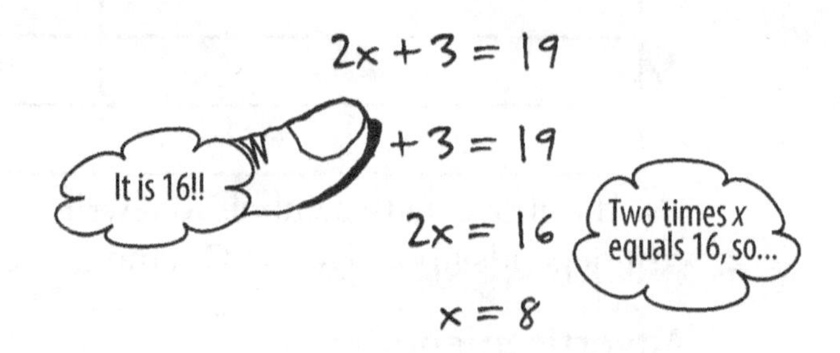

Then ask yourself: What number would I add to 3 to get 19? You would add 16. That means that 2 times the missing number (x) must equal 16. The missing number must be 8!

Does it work? Replace the "x" with an 8 to see whether it makes sense:

$$2(8) + 3 = 19$$

It works! Try this method with the problems below. Remember to rethink your questions when you are subtracting, not adding, a number.

Balance the amounts. For Questions 1–10, find the missing number using the cover-up method.

1. $2x + 10 = 20$ $x =$ ______

2. $5x + 9 = 49$ $x =$ ______

3. $2x - 50 = 150$ $x =$ ______

4. $4x - 5 = 15$ $x =$ ______

5. $4x + 20 = 100$ $x =$ ______

6. $10x + 900 = 1{,}000$ $x =$ ______

7. $100x - 50 = 150$ $x =$ ______

8. $8x + \frac{1}{2} = 56.5$ $x =$ ______

9. $\frac{x}{2} + 7 = 27$ $x =$ ______

10. $\frac{x}{10} + 8 = 88$ $x =$ ______

For Questions 11–15, write the equation first, and then solve it using the finger cover-up method.

11. If you double a certain number and add 3, you will get 7.

The equation is ______________________. $x =$ ______

12. Multiply a certain number by 5 and add 3 to get 28.

The equation is ______________________. $x =$ ______

13. If you multiply a certain number by 1,000 and add 250, you will get 6,250.

The equation is ______________________. $x =$ ______

14. Multiply a certain number by 7 and subtract 4 to get 59.

The equation is ______________________. $x =$ ______

15. Half a certain number minus 10 is 41. Find the number.

The equation is ______________________. $x =$ ______

Extension: Looking at Four Graphs

Juania says: "I really do not think it makes much difference for me. On Plans A, B, and D, I will pay about the same … But Plan C is definitely out."

What do you think her calling pattern is? Use the graph to support your conclusion.

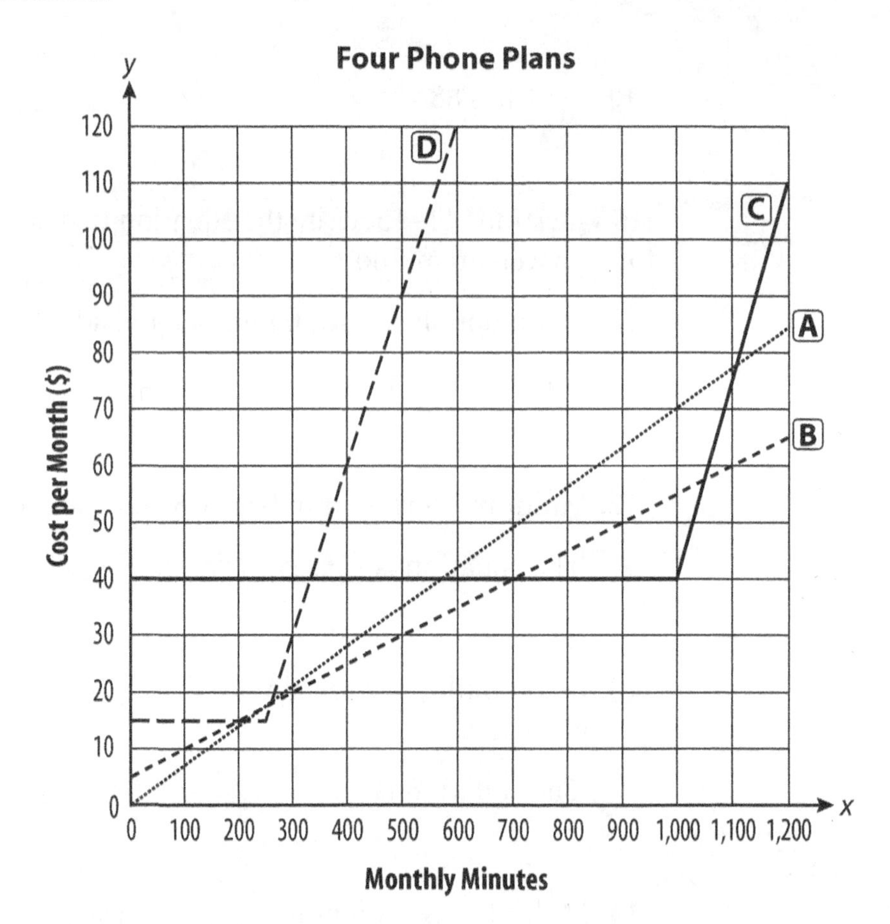

Test Practice

1. Which is a true statement?
   (1) $15 + 10 < 20$
   (2) $20 > 5 + 10$
   (3) $4(9) > 9(4)$
   (4) $4 + 9 > 9 + 4$
   (5) $20 + 5 < 15$

2. Which equation below would have the steepest graph?
   (1) $y = 2x + 20$
   (2) $y = 2x + 10$
   (3) $y = x + 1{,}000$
   (4) $y = 3x + 5$
   (5) $y = x - 1{,}000$

3. In which equation is 40 the value of x?
   (1) $x + 5 = 35$
   (2) $x - 5 = 45$
   (3) $45 = 2x - 5$
   (4) $x = 45 - 5$
   (5) $\frac{x}{2} = 90$

4. Which graph might represent the equation $y = 10$?

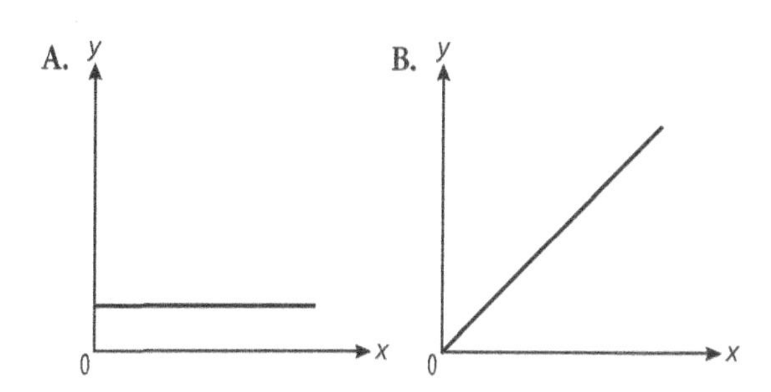

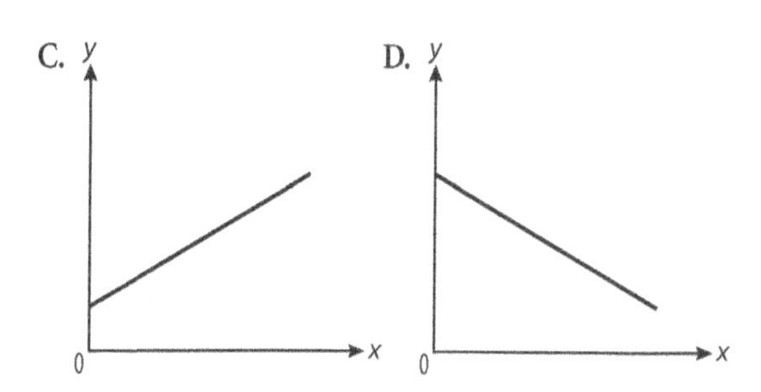

   (1) A
   (2) B
   (3) C
   (4) D
   (5) None of the above

5. Pedro likes to talk every day as much as he can on the phone. Which graph below represents a phone plan he would want? Cost of service is listed on the y-axis and minutes used on the phone is listed on the x-axis. The scales and intervals are the same on all graphs.

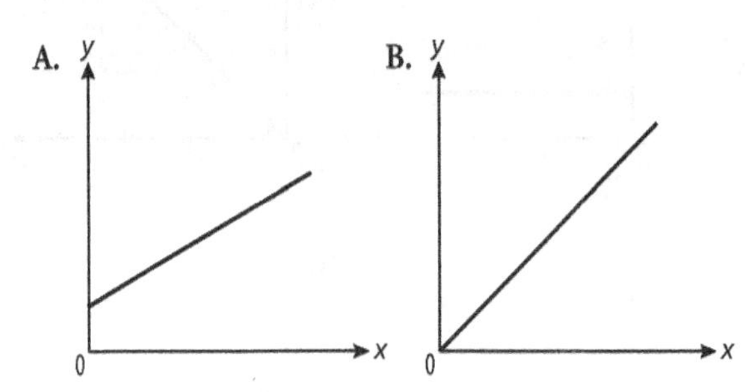

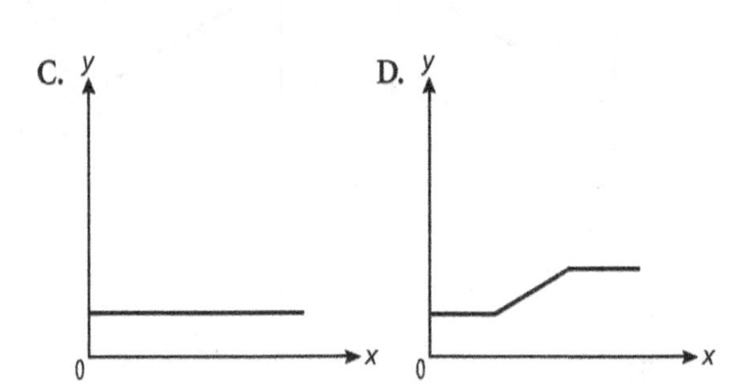

(1) A

(2) B

(3) C

(4) D

(5) None of the above

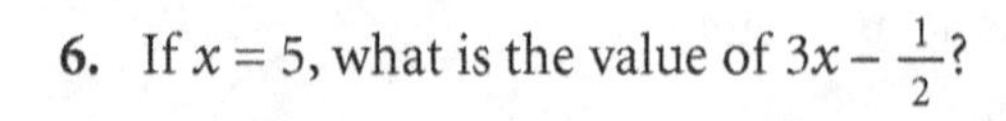

6. If $x = 5$, what is the value of $3x - \frac{1}{2}$?

LESSON 10

Signs of Change

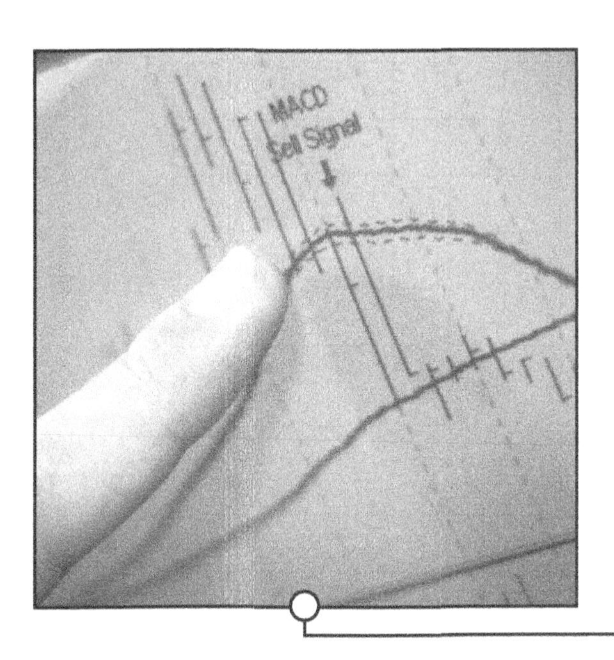

How do I know which one is changing fastest?

An important feature of linear relationships is that no matter how quickly or slowly the change happens, the *rate* at which the change takes place is constant.

Understanding the **rate of change** helps you get information more quickly from graphs, tables, and equations, and it gives you a good idea about how fast a pattern grows or shrinks.

In this lesson you look at patterns in a new way, and you will focus on how fast change happens. This is a different way to look at the relationship between x and y, as you will see.

Activity 1: Seeing the Constant Rate of Change

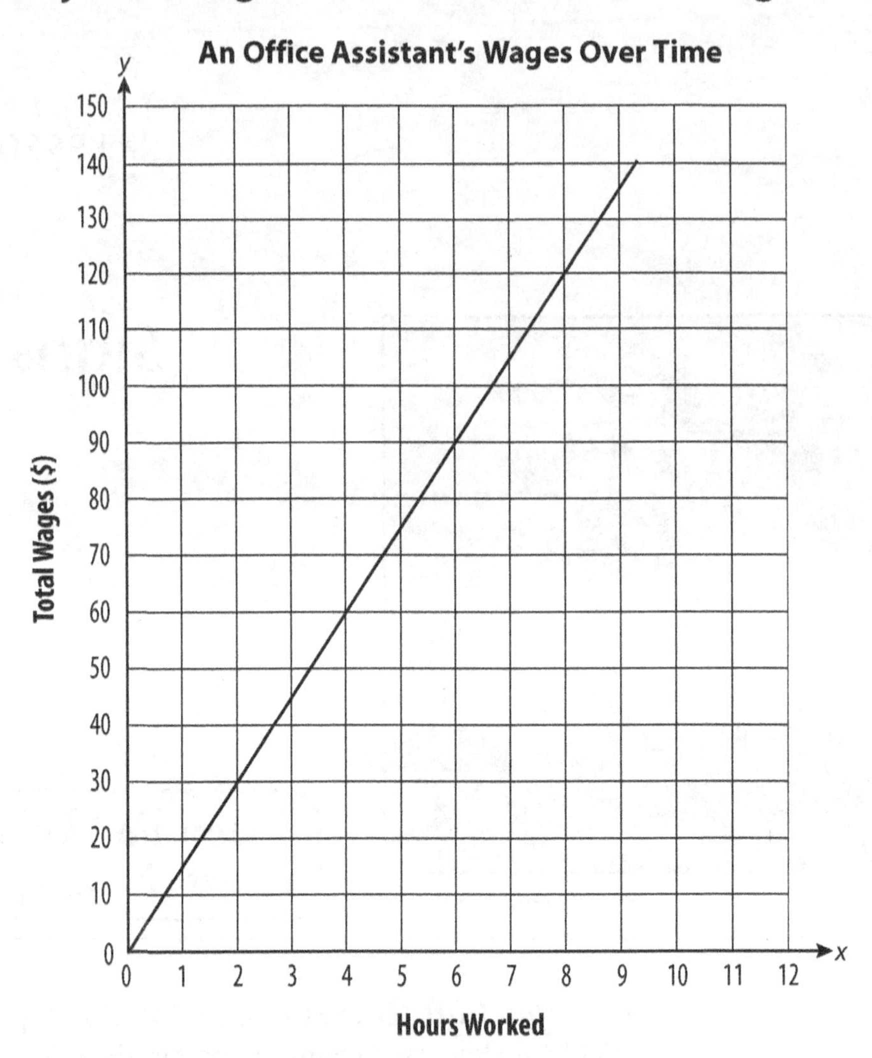

Looking at the Rate of Change in the Graph

1. Describe the rate of change. Tell what happens to the office assistant's wages over time. How fast are her wages accumulating?

2. Show on the graph where you see the rate of change.

3. Mathematically speaking, the rate of change is **constant** (always the same). How do you know it is constant?

What If...

4. If the office assistant received a pay raise to $18 per hour ($18/hr.), how would the graph look different? Sketch the two graphs to show the differences.

5. If the office assistant had received a $100 bonus for signing on with the company when she was making only $15/hr., how would that have changed the graph?

6. What makes the slant, or **slope**, of the graph steeper or flatter? Use examples to support your explanation.

Activity 2: Ranking Rates of Change

Your teacher will give you a set of tables, graphs, and equations. Each representation you see communicates a different situation, but you can compare the constant rates of change for each of them. Rank each representation in order from the slowest growing or smallest rate of change (1) to the fastest growing or largest rate of change (6).

1. How we ranked the representations from slowest rate of change to fastest rate of change:

   (1) ______________________________

   (2) ______________________________

   (3) ______________________________

   (4) ______________________________

   (5) ______________________________

   (6) ______________________________

2. Explain how you made your decisions and what you discovered. Tell what you looked at when making a decision.

3. Choose one representation. What is a real situation that could be explained by that representation?

Practice: Watching Money Grow in a Table

You invested $1,500. You decided to put $500 in three certificates of deposit (CDs).

1. Use the table to describe the rate of change (in dollars per quarter) in each of these CDs.

Quarter	Value		
	Eastbank	U.S. Savings	People's Bank
0	$500	$500	$ 500
1	$525	$550	$ 585
2	$550	$600	$ 670
3	$575	$650	$ 755
4	$600	$700	$ 840
5	$625	$750	$ 925
6	$650	$800	$1,010
7	$675	$850	$1,095
8	$700	$900	$1,180

2. What do you think the graph for each of these CDs would look like?

3. Carefully draw the graph on grid paper. What do you notice?

4. What will happen to the three lines if you extend the graph to 10 years? Why?

5. How do the equations for each CD differ?

Practice: Whose Total Earnings Change the Fastest?

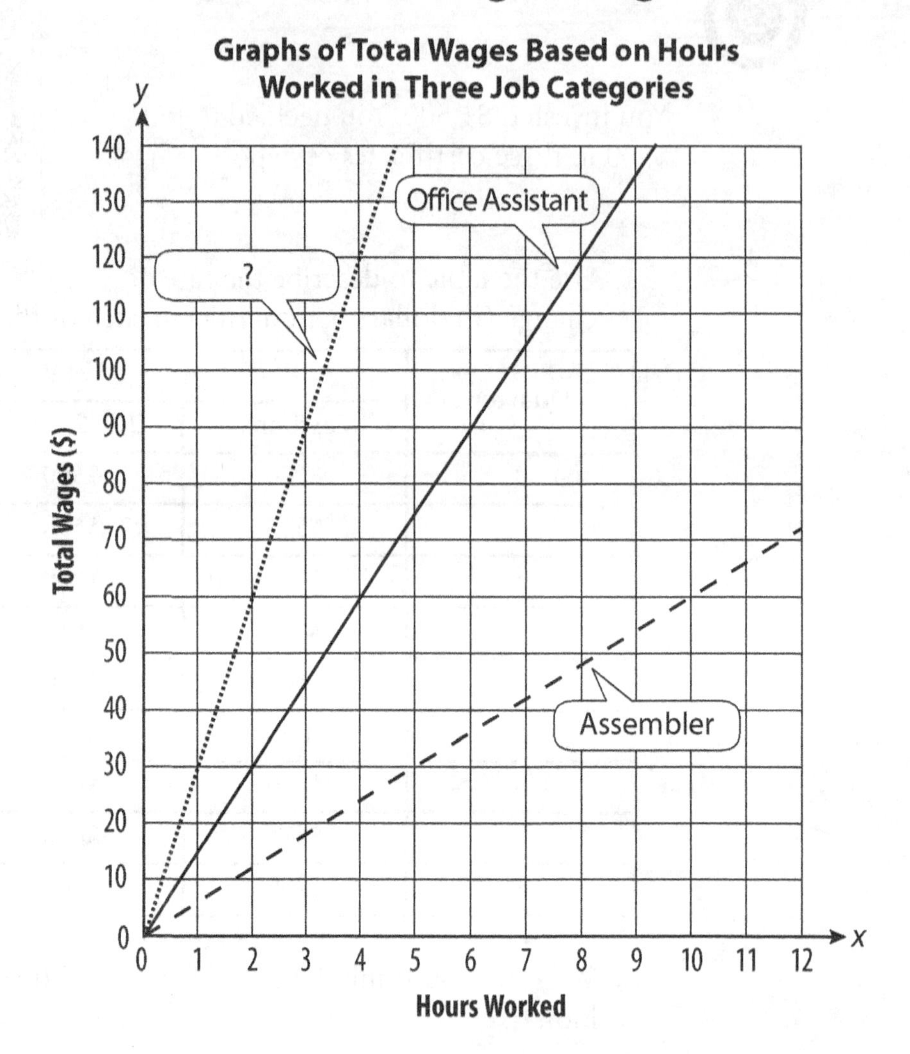

1. Make several observations about the graphs.

2. What type of job do you think the mystery job might be?

3. Show the rate of change in each graph.

4. Whose total earnings show the fastest constant rate of change? How do you know?

5. In what ways do the equations for each job category differ?

6. If this graph showed earnings for a year, what would happen on the graph? Why? What does that tell you about the differences in total wages for the three jobs as time passes?

Symbol Sense Practice: Solving More Two-Step Equations

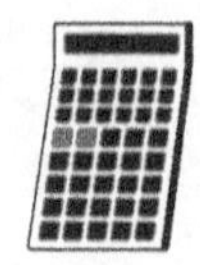

You have solved one- and two-step equations in earlier *Symbol Sense Practices*. To solve the equations below, you can use either of two different methods. Sometimes one makes more sense than the other to use. You decide!

Example: $8(x + 7) = 800$

First Way:

$8(x + 7) = 800$

$8(\quad) = 800$

Eight times some number equals 800. I am covering 100!

$(x + 7) = 100$

What number plus 7 equals 100?

$x = 93$

Second Way:

$8(x + 7) = 800$

$8x + 56 = 800$

$\quad + 56 = 800$

I am covering 744!!

$8x = 744$

What number times 8 equals 744?

$x = 93$

Find the value of x in each of the equations below.

1. $6(x + 2) = 54$ $\qquad x =$ ________

2. $12(x + 1) = 144$ $\qquad x =$ ________

3. $6(x - 3) = 120$ $\qquad x =$ ________

4. $11(x + 3) = 1{,}111$ $\qquad x =$ ________

5. $9(x - 8) = 63$ $\qquad x =$ ________

6. $9x - 8 = 64$ $\qquad x =$ ________

7. $15(x - 12) = 180$ $\qquad x =$ ________

8. $15x - 12 = 168$ $\qquad x =$ ________

Test Practice

Table A

x	y
3	22
4	29
5	36

1. A rule for Table A could be written as

(1) $y = x + 7$

(2) $y = \frac{x}{3} - 1$

(3) $y = 7x + 1$

(4) $y = 7x$

(5) $y = 3x + 1$

2. According to the rule in Table A, if $x = 15$, the value for y would be

(1) 22

(2) 36

(3) 105

(4) 106

(5) 306

3. The rate of change in Table A could be described this way:

(1) as x increases by 1, y increases by 7

(2) as x decreases by 1, y increases by 7 plus 1

(3) as x increases by 1, y decreases by 7

(4) as x increases by 1, y increases by 17

(5) as x increases by 7, y increases by 1

4. Which equation shows the fastest rate of change?

(1) $y = 10x$

(2) $y = 60 + x$

(3) $y = 8x + 750$

(4) $y = 6x - 5$

(5) $y = \frac{x}{6}$

5. $10(x + 200) = 2{,}500$. Solve for x.

(1) $x = 25$

(2) $x = 50$

(3) $x = 100$

(4) $x = 250$

(5) $x = 500$

6. Choose the one point you would find in a graph for Table B, and plot it on the graph provided.

Table B

x	y
5	16
6	19
7	22

(1) (0, 0)

(2) (0, 1)

(3) (1, 6)

(4) (2, 6)

(5) (3, 6)

y

6
5
4
3
2
1
-6 -5 -4 -3 -2 -1 0 1 2 3 4 5 6 x
-1
-2
-3
-4
-5
-6

LESSON 11

Rising Gas Prices

Are gas prices rising steadily?

Rate of change is an important concept in finance, science, and politics.

How fast does your money grow when you put it in a savings account? How fast do bacteria multiply under certain conditions? How long will it take for the radioactivity to decrease in a contaminated area? Looking at rate of change in patterns will help you answer these kinds of questions.

In this lesson, you will use tables, graphs, and equations to see where rate of change is constant and where it is not.

Activity 1: Patterns of Change

Examine and compare these patterns.

Pattern 1

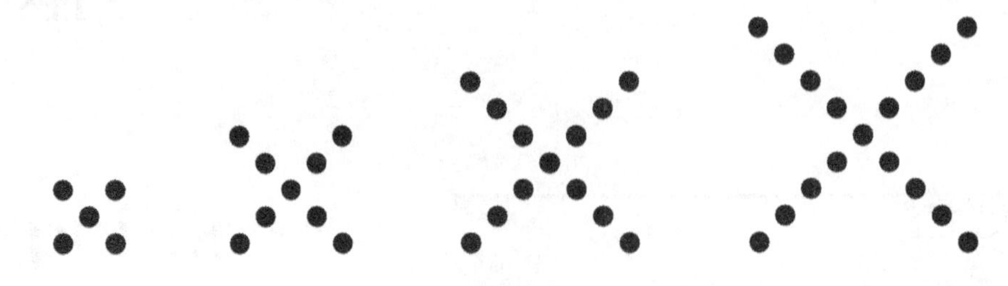

Pattern 2

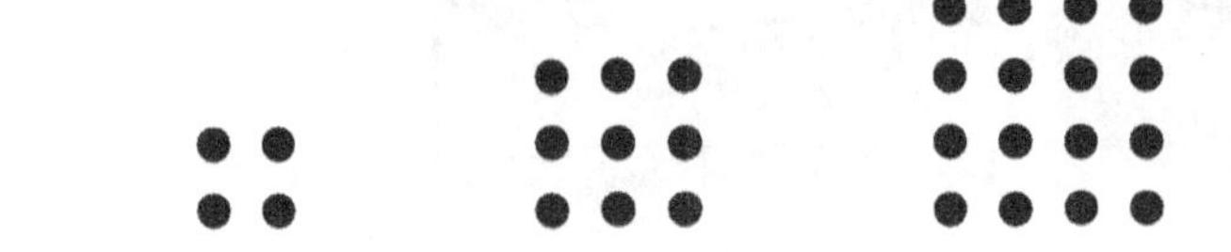

1. Create a *table* and a *graph* for each pattern. Draw both graphs on the same grid.
2. Which one is a linear pattern (has a constant rate of change)? How do you know?
3. How many dots will be in the 10th, 25th and *n*th entry of each pattern?

> The *n*th entry refers to *any* case. It tells the general rule to find the Output when the Input number is *n*.

4. Which grows faster? Why?

Patterns of Change

Case Number	Number of Dots Pattern 1	Number of Dots Pattern 2
1		
2		
3		
4		
5		
10		
25		
n		

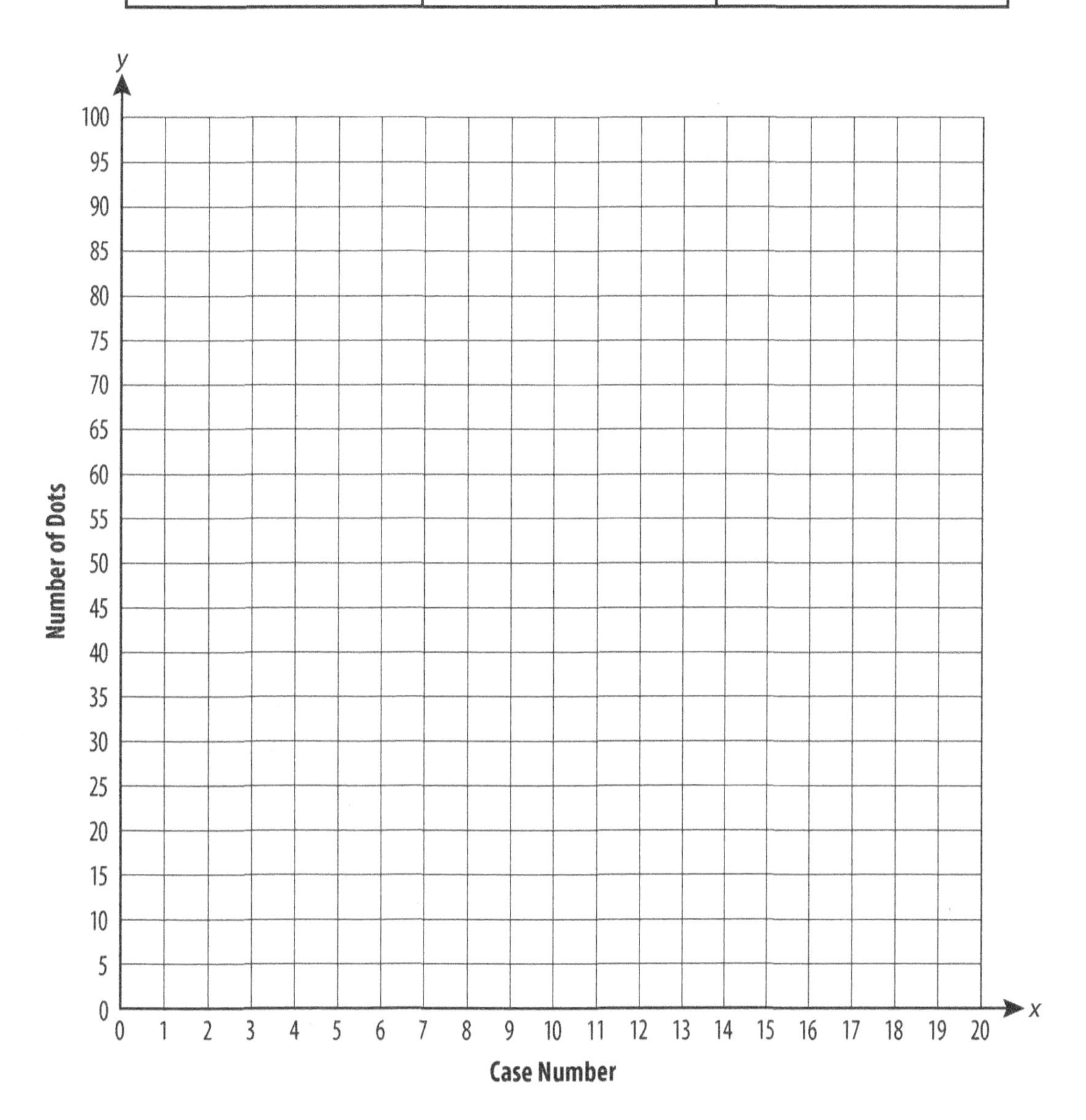

Activity 2: Sky-High Gas Prices

Joe's Gas Station and Metric Service Station used to charge the same price for gas—$1.50/gallon. Six months ago things changed. Joe and Metric started increasing their prices like crazy.

At Joe's:
Every month for the last six months, the price of gas has gone up $2.00.

At Metric:
Every month for the last six months, the price of gas has doubled from what it was the month before.

1. Make a *table* and draw a *graph* for each of these situations. Draw both graphs on the same piece of grid paper.

2. Which gas price is changing at a constant rate? How do you know?

3. At which month would you have started to get suspicious that one gas station's price was really sky rocketing? Why?

4. If this pattern continues, how much will gas cost at each station by the end of a year? Explain how you figured this out.

Sky-High Gas Prices

Month Number	Price per Gallon at Joe's	Price per Gallon at Metric
0		
1		
2		
3		
4		
5		
6		
12		
n		

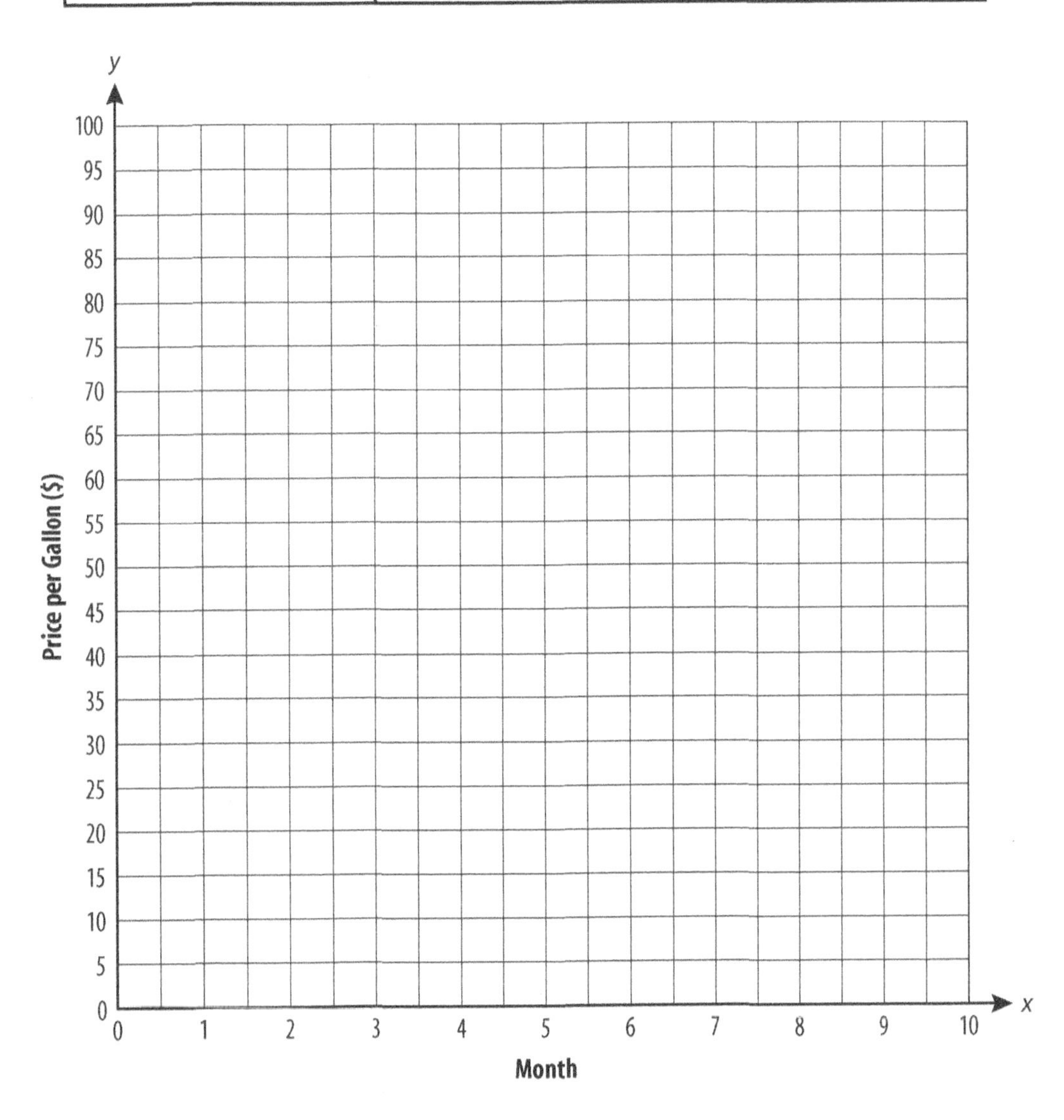

Practice: More Patterns of Change

Examine these five patterns. Which change at a constant rate? Which do not? Graph one of the linear patterns and one of the **nonlinear patterns.**

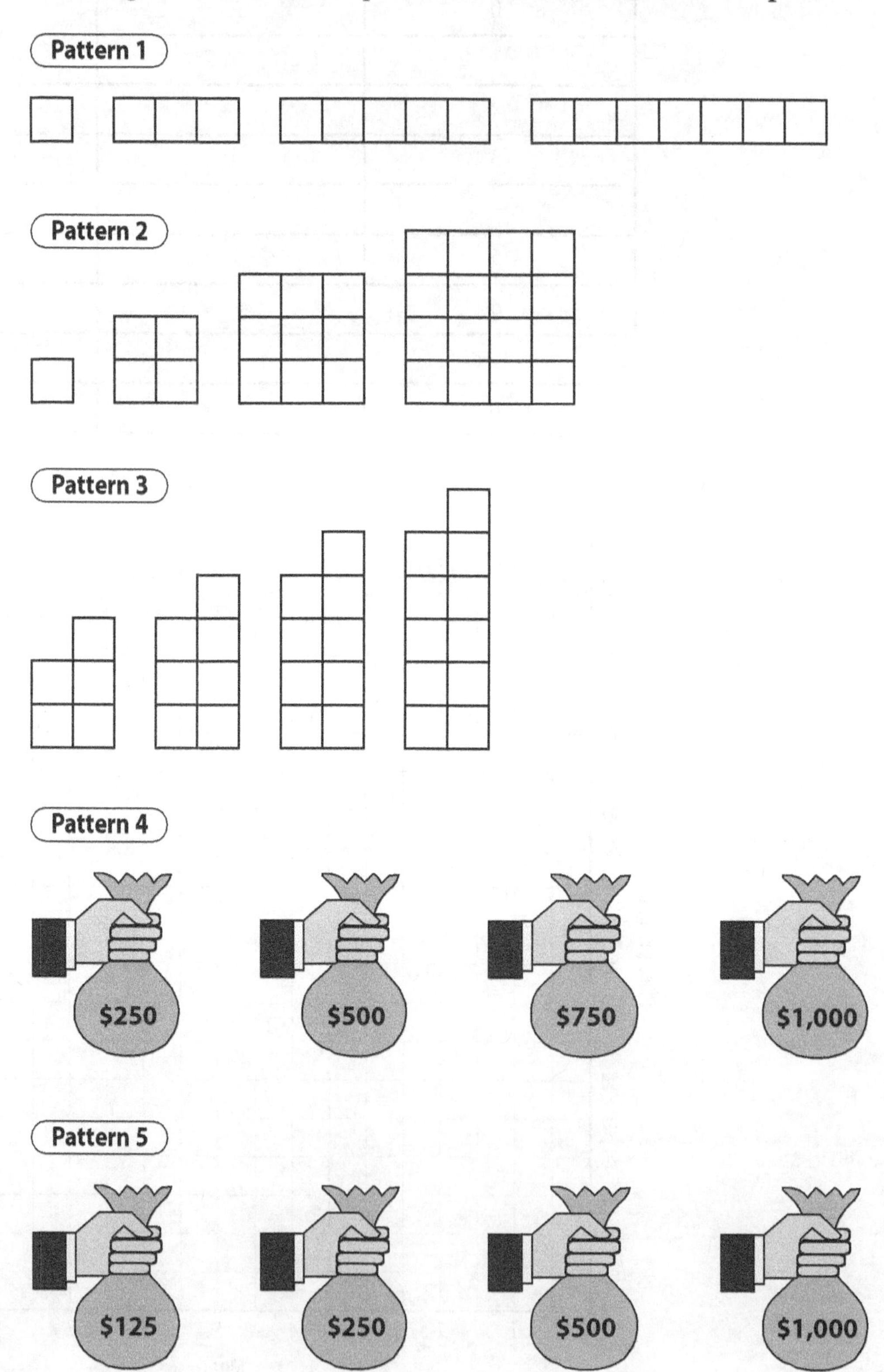

Practice: Investigation A—Expanding Squares

Perimeter is the distance *around* a shape. Area is the number of square units *covered by* a shape.

Here is a set of eight squares with sides of 1 unit, 2 units, 3 units, … up to 8 units.

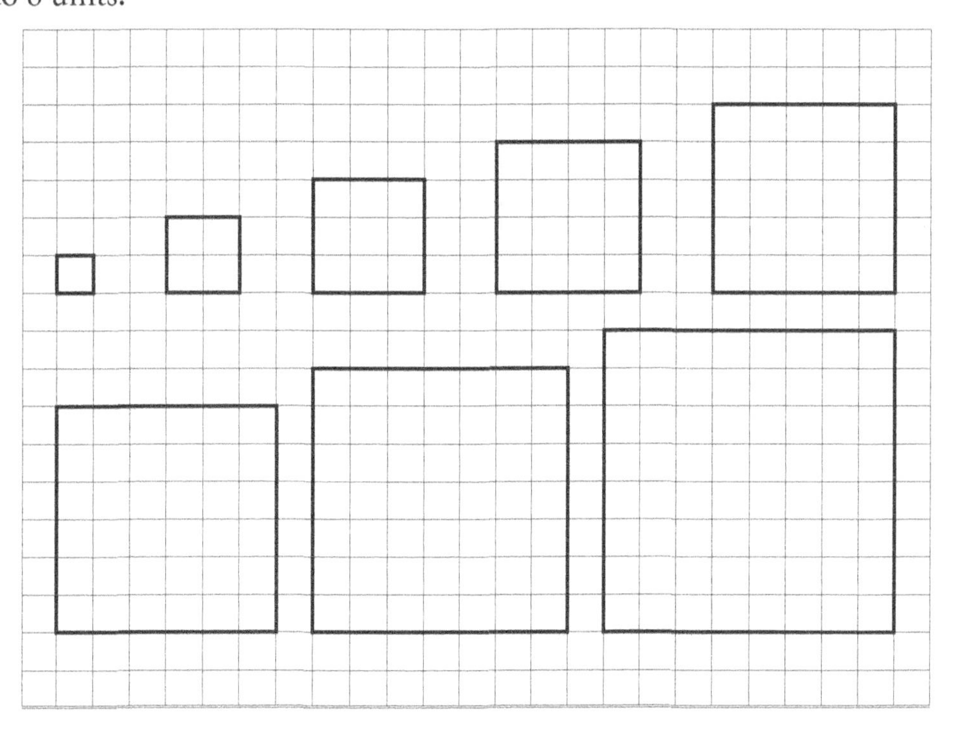

1. Determine the perimeter of each square.
2. Determine the area of each square.
3. What would the perimeter be for a square with a side of 15 units? What would the area be?
4. When you lengthen the side of a square by one unit, what happens to the perimeter? What happens to the area?
5. Compare the rates of growth for the area and the perimeter. Which grows faster? Why?
6. Which grows at a constant rate—the perimeter or the area? Show your answer on a graph.

Practice: Investigation B—What Is Behind the Door?

You are on a game show, and you may choose one of two doors.

Mystery Door 1:
"Open me, and I will give you 10 dollars on the first day. Each day after that (for a total of 10 days), I will double the amount I gave you the day before."

Mystery Door 2:
"Open me, and I will give you $100 more each day for 10 days."

1. Which door would you pick? Why?

2. Use a graph and a table to support your choice.

3. Which of the totals grows at a constant rate? How do you know?

Practice: Investigation C—What Size Is That Copy?

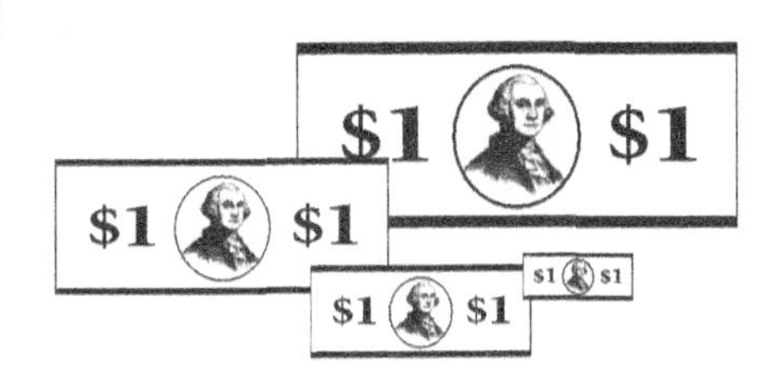

On a copy machine, make four copies of a dollar bill, with the different enlargement/reduction settings shown in the table below. Take measurements of the length, width, and area of each of the reproductions, and record the measurements in the table. Then draw a graph for the length, width, and area growth. Put the percent of the enlargements on the x-axis.

% Enlargement/ Reduction	Length	Width	Area
50			
100			
150			
200			

1. For each time you increase the enlargement setting by 50%, how do the length, the width, and the area change?

2. Does the length grow at a constant rate? Demonstrate with a graph.

3. Does the area grow at a constant rate? Demonstrate with a graph.

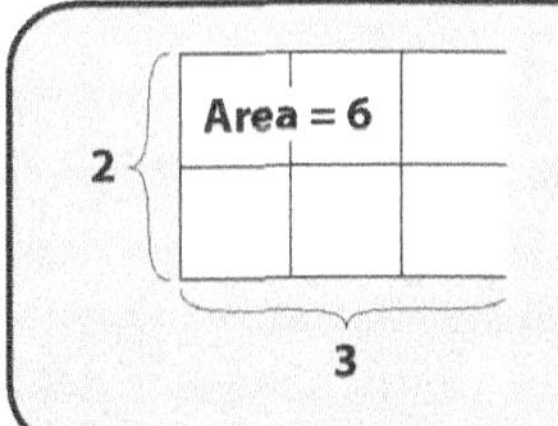

To find the area of a rectangle, multiply the length by the width.

Symbol Sense Practice: Squaring Numbers

Squaring is not the same as doubling.

Squaring is different.

You double a number by multiplying the number by 2: $2(n)$

You square a number by multiplying it by itself: $(n)(n) = n^2$.

The small raised number is called an **exponent**.

1. Fill in this table of common doubles and squares. How is the $2n$ pattern different from the n^2 pattern?

n	$2n$	n^2
0		
1		
2		
3		
4		
5		
6	12	
7		
8		
9		81
10		
11		
12		

2. Use a calculator to fill in this table of more unusual doubles and squares.

n	$2n$	n^2
2.5		
	100	
		10,000
20		

Symbol Sense Practice: Using Exponents

Fill in the table below.

Exponent Form	In Words	As Repeated Multiplication	Value
10^4	ten to the fourth power	10 x 10 x 10 x 10	10,000
5^2	five squared	(5)(5)	25
2^5			
		(4)(4)(4)	
		(3)(3)(3)(3)	
	eight to the third power, or eight cubed		
	ten cubed, or ten to the third power		
16^2			
0^4			
1^4			

Test Practice

1. Which of these patterns is linear (shows a constant rate of change)?

   A. 1, 3, 5, 7 **B.** 2, 4, 6, 8

   C. 1, 4, 9, 16 **D.** 2, 9, 16, 23

   (1) A only

   (2) B only

   (3) A and B only

   (4) A, B, and D only

   (5) A, B, C, and D

2. In the story, "A Grain of Rice," a young man cures a princess and takes as his prize a grain of rice doubled every day for 100 days. After how many days will he have 128 grains of rice? (On day one, he has 1 grain.)

   (1) 4

   (2) 5

   (3) 7

   (4) 8

   (5) 9

3. A picture of 3^2 might look like this:

   (1)

   (2)

   (3)

   (4)

   (5)

4. If $n^2 = 144$, what is the value of n?

   (1) 12

   (2) 14

   (3) 72

   (4) 140

   (5) 144

5. When does $2x = x^2$?

   (1) Never

   (2) Always

   (3) When $x = 0$ and when $x = 2$

   (4) When $x = 1$

   (5) When $x = 4$

6. Plot the point for $y = x^2 - 10$, when $x = 4$.

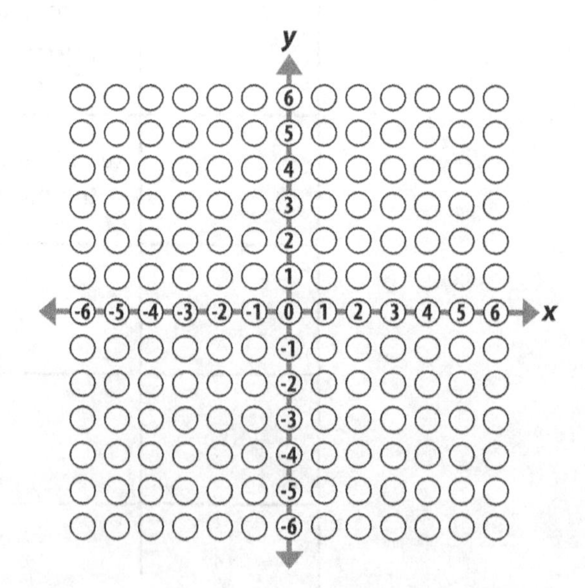

LESSON 12

The Patio Project

How can you predict what you will need for the larger design?

In this final lesson, you will build patio designs that involve three kinds of tile patterns. The patio design provides a model to understand how different patterns grow.

Some patterns change at a constant rate, and others do not. You will create formulas to figure out how many of each tile you will need for any size patio.

Activity: The Patio Project

Imagine that you manufacture patio tiles for do-it-yourself projects. For square patios, you have created a design that uses solid-color tiles in the center, striped tiles in the corners, and white tiles along the sides or edges.

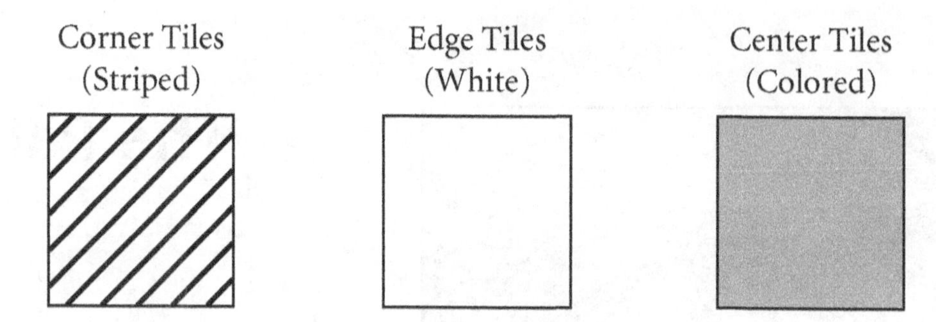

For example, a four-by-four patio would look like this:

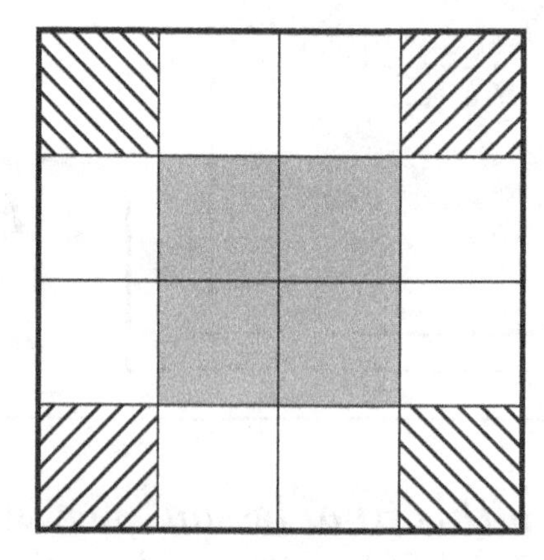

To supply customers with the correct amount of tiles for their projects, you developed the following table. Before you completed the table, a customer arrived and asked about a special project.

Patio Size	Number of *Corner* Tiles (Striped)	Number of *Edge* Tiles (White)	Number of *Center* Tiles (Colored)	Total Number of Tiles
2 by 2				
3 by 3				
4 by 4				
5 by 5				
6 by 6				

The Customer's Question

The customer posed this question: "What is the total number of each type of tile I will need to build a 100-by-100 patio?"

Explain how you could determine the number of each tile needed for this size patio, so the customer trusts your advice. Prepare a presentation that includes diagrams, graphs, and/or equations, as well as your explanation to the customer. Use the grid below if necessary.

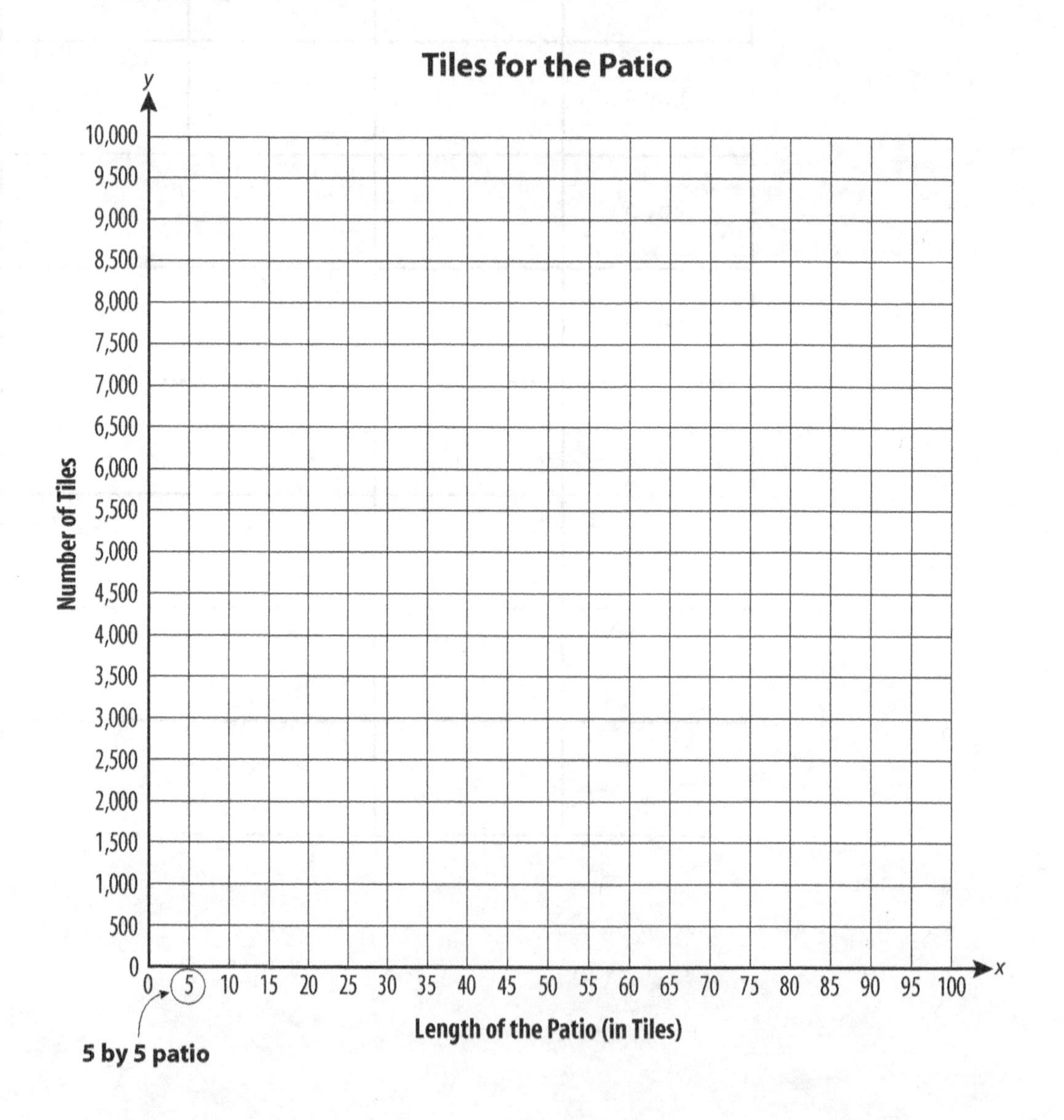

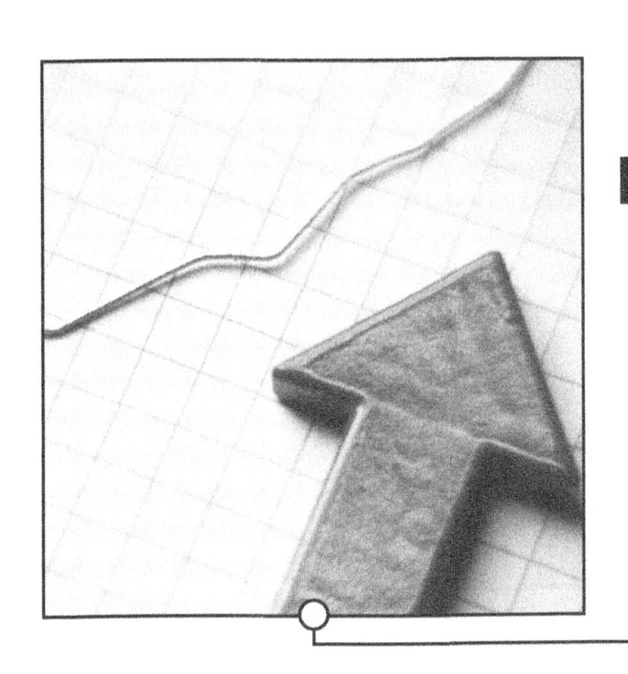

Closing the Unit: Putting It All Together

Where do I go from here?

This unit on algebraic thinking is almost finished, and now you have a chance to pull together some of the main ideas that were covered.

For the first activity, your teacher will give you the *Final Assessment*. There are three tasks to be done, in this order:

1. *Some Personal Patterns*: Pick a personal pattern and represent it with a table, a graph, a rule in words, and an equation.
2. *Match Situations with Representations*: Match situations with graphs, tables, and equations, and then solve three problems.
3. *Symbol Sense Review*: Review symbolic notation practices.

The other activity is the *Mind Map* and *Unit Reflection*. What does the word "algebra" bring to mind? What did you learn that you think is important? What lessons do you remember most vividly?

Activity 2: Mind Map and Unit Reflection

Mind Map

Make a Mind Map showing math words and ideas that come to mind when you think of *algebra*. Remember to link together ideas and words that are related.

Unit Reflection

1. What have you learned?
   Recall all the work you have done in *Seeking Patterns, Building Rules*. List or describe what you have learned about algebra that you think is important. Why is it important to you?

2. What lessons do you remember most vividly?
   Review the *Contents*, page iii, and consider which lessons you remember most clearly and why. Describe your experiences below.

VOCABULARY

Lesson	Terms, Symbols, Concepts	Definitions and Examples
Opening the Unit	algebra	
	pattern	
1	equation	
	In-Out table	
	table	
	variable	
4	coordinate graph	
	increments	
	inverse operations	
	labeling points	
	origin	
	x-axis	
	y-axis	

VOCABULARY *(continued)*

Lesson	Terms, Symbols, Concepts	Definitions and Examples
6	circumference	
	diameter	
	formula	
	pi	
8	point of intersection	
	y-intercept	
9	flat-line graph	
	line steepness	
10	constant rate of change	
	linear relationship	
	slope	
11	exponent	
	exponential relationship	
	nonlinear rate of change	
	nonlinear pattern	

VOCABULARY *(continued)*

Lesson	Terms, Symbols, Concepts	Definitions and Examples

REFLECTIONS

OPENING THE UNIT: Seeking Patterns, Building Rules

What patterns were discussed in class today?

How do you think understanding patterns will help you in your own life?

LESSON 1: Guess My Rule

In this lesson you did a lot of work with In-Out tables. What hints about looking for patterns in tables do you want to remember?

LESSON 2: Banquet Tables

What did you learn by working on the banquet tables problem? How did a diagram, table, rule, or equation help you predict?

LESSON 3: Body at Work—Tables and Rules

What is important to remember about how to make connections between a table and a rule?

LESSON 4: Body at Work—Graphing the Information

Make a Mind Map of the word "graph" to help you remember all the terms you learned today. How does a graph connect with a table and an equation?

LESSON 5: Body at Work—Pushing It to the Max

For different situations, the graph slants in different directions. What two ways have you seen line graphs slant? What caused them to slant in different ways?

LESSON 6: Circle Patterns

Draw a picture of a circle and mark its diameter. Label the circumference and diameter. Explain one of the formulas you worked with today.

MIDPOINT ASSESSMENT

Reflect on your experiences in class today. What seemed easy and what seemed difficult?

LESSON 7: What Is the Message?

Math equations communicate information. What information do the two equations $y = x + 5$ and $y = 5x$ tell you?

LESSON 8: Job Offers

One of the job-offer line graphs started at the origin, and one started high on the y-axis. Why?

Both line graphs intersected at one point. Why?

LESSON 9: Phone Plans

You examined tables, graphs, and equations for various phone plans today, and then you matched them. List the parts of the tables, graphs, and equations you used to match the phone plans.

LESSON 10: Signs of Change

How can you tell which pattern has a faster rate of change when looking at two tables, two graphs, or two equations?

LESSON 11: Rising Gas Prices

Linear and nonlinear patterns look different when you see them in a table or graph them. Explain why, using diagrams and words.

LESSON 12: The Patio Project

Think about how you and your partner solved the patio projects problem today. What worked well; what did not work well? Faced with a similar problem, what would you do the same, and what would you do differently?

LESSON 12: The Ratio Project

[illegible]